International Series on Computer Entertainment and Media Technology

Series Editor
Newton Lee
Department of Applied Computer Science
Woodbury University
Burbank, CA, USA

The International Series on Computer Entertainment and Media Technology presents forward-looking ideas, cutting-edge research, and in-depth case studies across a wide spectrum of entertainment and media technology. The series covers a range of content from professional to academic. Entertainment Technology includes computer games, electronic toys, scenery fabrication, theatrical property, costume, lighting, sound, video, music, show control, animation, animatronics, interactive environments, computer simulation, visual effects, augmented reality, and virtual reality. Media Technology includes art media, print media, digital media, electronic media, big data, asset management, signal processing, data recording, data storage, data transmission, media psychology, wearable devices, robotics, and physical computing.

More information about this series at http://www.springer.com/series/13820

Duncan Williams • Newton Lee
Editors

Emotion in Video Game Soundtracking

Editors
Duncan Williams
Digital Creativity Labs
University of York
York, UK

Newton Lee
Department of Applied Computer Science
Woodbury University
Burbank, CA, USA

ISSN 2364-947X ISSN 2364-9488 (electronic)
International Series on Computer Entertainment and Media Technology
ISBN 978-3-319-89165-1 ISBN 978-3-319-72272-6 (eBook)
https://doi.org/10.1007/978-3-319-72272-6

Printed on acid-free paper

This Springer imprint is published by Springer Nature
The registered company is Springer International Publishing AG
The registered company address is: Gewerbestrasse 11, 6330 Cham, Switzerland

Contents

About the Editors

Newton Lee is CEO of Newton Lee Laboratories, LLC; president of the Institute for Education, Research, and Scholarships; adjunct professor at Woodbury University School of Media, Culture & Design; and editor-in-chief of Association for Computing Machinery Computers in Entertainment. Previously, Lee was a computer scientist at AT&T Bell Laboratories, senior producer and engineer at the Walt Disney Company, and research staff member at the Institute for Defense Analyses. He was the founder of Disney Online Technology Forum, creator of Bell Labs' first-ever commercial AI tool, and inventor of the world's first annotated multimedia OPAC for the U.S. National Agricultural Library. Lee graduated Summa Cum Laude from Virginia Tech with a B.S. and M.S. degree in Computer Science, and he earned a perfect GPA from Vincennes University with an A.S. degree in Electrical Engineering and an honorary doctorate in Computer Science.

Duncan Williams is a Researcher in the Digital Creativity Labs at the University of York, UK. He holds a PhD in Signal Processing and Psychoacoustics from the University of Surrey, UK, and has been a psychoacoustic consultant for Honda Automotive, the British Broadcasting Corporation, and the UK Government. Duncan is a Fellow of the Higher Education Academy and was nominated by the students of Plymouth University in 2016 for an "Inspirational Teaching" award. He also holds teaching qualifications from Surrey University (2004), Apple Inc. (2007), and Plymouth University (2015). His catalogue of music-for-picture includes a published back catalog with EMI, Sony/ATV, and DeWolfe. Duncan is an editorial board member of Musicae Scientae (Sage), the Journal of Creative Music Systems (University of Huddersfield Press), and the Encyclopedia of Computer Graphics and Games (Springer). A blog and publications list is updated regularly at www.duncanwilliams.info.

About the Authors

David Bessell has contributed a chapter titled "An auto-ethnographic approach to creating the emotional content of horror game soundtracking."

The original meaning of melodrama is music and drama. This chapter explores how the video game context in relation to non-linear musical form impacts on the ability to generate appropriate emotions at appropriate points in gameplay, particularly relevant to horror soundtracking that relies heavily on music to augment emotional impact.

Dr. Dave Bessell has been active in the field of popular music for many years. He also studied classical composition and orchestration at the Royal College of Music, jazz guitar with John Etheridge, holds a doctorate in Music, and currently teaches Music and Music Technology at Plymouth University. He can be found performing on guitar or electronics from time to time in a variety of styles. He was one of the pioneering writers on the subject of video game soundtracking, and he is currently working on soundtrack for a horror game in production.

Giles Hooper (Senior Lecturer in Music, University of Liverpool) has contributed a new chapter on the role and function as well as the potential challenges of music and sound in a virtual reality context taking a focus on the use of music in cut scenes.

Giles Hooper completed his PhD, The study of music and the status of musical knowledge, at the University of Keele in 2003. After teaching at Keele, Exeter, and Bristol, he was appointed as a lecturer in the School of Music in 2005. Hooper's work is currently in wide-ranging research interests including twentieth-century music, critical theory, and analysis. In 2010, Hooper was appointed Head of the School of Music. His best known publication is *The Discourse of Musicology* published in 2006. https://en.wikipedia.org/wiki/Giles_Hooper

Alexis Kirke presents a chapter on the emerging field of "audio-only" games. There has been an increasing popularity in audio-only games, particularly for mobile devices. These are usually based on spoken-word and sound-effect scene setting. This chapter investigates whether this can be taken further into games, which are entirely abstract sonic and music experiences, and presents a first example based on Minecraft: Musicraft.

Dr. Alexis Kirke is a Senior Research Fellow in Plymouth University's Interdisciplinary Centre for Computer Music Research. He publishes in AI and HCI and is also a well-known interdisciplinary composer and artist.

Damian T. Murphy is Professor of Sound and Music Computing, Department of Electronics, University of York, where he has been a member of academic staff since 2000, and the University Research Champion for Creativity. His research focuses on virtual acoustics, spatial audio, physical modelling, and audio signal processing. He has been principal investigator on a number of related AHRC and EPSRC funded projects relating to room acoustics simulation and auralisation and published over 100 journal articles, conference papers, and books in the area. He is a member of the Audio Engineering Society, a Fellow of the Higher Education Academy, and a visiting lecturer to the Department of Speech, Music and Hearing at KTH, Stockholm, where he specialises in spatial audio and acoustics. He has also held visiting researcher status at a number of universities internationally. Dr. Murphy is an active sound artist and in 2004 was appointed as one of the UK's first AHRC/ACE Arts and Science Research Fellows, investigating the compositional and aesthetic aspects of sound spatialisation, acoustic modelling techniques, and the acoustics of heritage spaces. His work has been presented in galleries nationally and at festivals and venues internationally and included varied collaborations with writers, photographers, and interactive digital artists. He is a founding member of Geodesic Arts through which most of his more recent work has been produced.

Joseph Rees-Jones is a PhD candidate working with Dr. Murphy at the University of York in the Department of Electronics. He is an active member of the audio engineering society working in the York Audio Lab. He holds an MSc by Research in Music Technology, having conducted extensive research investigating the implementation of spatial audio in video games.

Federico Visi (Post-doctoral Research Fellow at the Universität Hamburg, Germany) has contributed a chapter that focuses on how motion capture can be used to control procedural audio and real-time signal processing synthesis in particular in extracting motion features from various devices and using them as an input to generative algorithms.

Federico Visi is a composer, producer, and sound designer. After obtaining his master's degree in communication, multimedia, and design, he studied music for image in Milan and composition at the music academy Accademia Pianistica in Imola. His studies and research focus on the interaction between sound, image, and motion and on creating meaning through the combination of multiple means of expression. He composed music and designed the sound for several films, documentaries, installations, and commercials and performed live in solo sets, with bands and in contemporary theatre and dance performances. http://www.federicovisi.com/bio/

Chapter 1
Welcome and Introduction from the Editors

Duncan Williams and Newton Lee

Welcome

Welcome to Emotion in Video Game Soundtracking. This book presents an overview of this exciting, emerging field. Before we begin, we should probably get one thing settled: is it "Video game" or "videogame" (or indeed, some other permutation)? According to industry insiders, the accepted standard was "videogame" until "video game" started to yield more internet search results (On "Videogame" Versus "Video Game" 2017). So, we are going with video game/s. But if this really worries you, feel free to substitute your preference each and every time you see an instance, which will be a lot by the time you have reached the end of this book.

Now that we've got that out of the way, regardless of your views, it is undeniable that the video games industry has become one of the biggest global markets in entertainment. Video game soundtracking has, by the same token, become one of the biggest outlets for commercial music, both in terms of composition, production, and dissemination. This book is specifically concerned with one aspect of soundtracking: that of emotion. This might be in the player, in the composer, the game designer, or the casual viewer, and this book will address each of these perspectives with practical and theoretical examples from each use-case.

Why is the focus of the book on emotion? The emotional impact of music has been well-documented, particularly when used to enhance the impact of a multi-modal experience, such as combining images with audio as found in the film industry, and, of course, a huge amount of the games industry. Soundtracking videogames presents a unique challenge compared to traditional sound-for-picture (i.e., film and

D. Williams (✉)
Digital Creativity Labs, University of York, York, UK
e-mail: duncan.williams@york.ac.uk

N. Lee
Department of Applied Computer Science, Woodbury University, Burbank, CA, USA

D. Williams, N. Lee (eds.), *Emotion in Video Game Soundtracking*, International Series on Computer Entertainment and Media Technology,
https://doi.org/10.1007/978-3-319-72272-6_1

television soundtracking) in that the narrative of gameplay is non-linear —Player dependent actions can change the narrative and thus the emotional characteristics required in the soundtrack. Is your player alive, dead, happy, sad, afraid, or angry? Each requires a different approach to the soundtrack in terms of emotional inference.

Some chapters in this book outline historical approaches to emotion measurement, whilst others consider how such measurement might be used to map musical features and soundtrack selections in the context of video game soundtracking. A number of cutting edge examples are given, including the use of biophysiological sensing (brainwaves, motion capture suits, and heart-rate sensors), as well as systems where the soundtrack is the game. We will hear from real-world game composers and consider next-generation examples which use automated repurposing of existing music (for example from a player's own library). Strategies for evaluating the success of video game music in this regard can be borrowed from music psychology, and combined with recent advances in computer science in order to derive next-generation applications for emotionally congruent videogame soundtrack generation. We also examine some implementations, including emotionally driven signal processing, speech synthesis, and the emerging field of affectively-driven algorithmic composition (ADAC or AAC). We predict that in the next ten years the videogame soundtracking landscape may be unrecognizable from the current state-of-the-art. We hope that this book thus presents a timely opportunity to both review the past and look to the future.

Who Is this Book for?

If you are reading this book, we expect you might be either a working, or aspiring professional in the video game industry, a scholar focused on interdisciplinary practice or branching out (e.g., from musicology, multimedia studies, film analysis, story-telling), or perhaps with an interest in the psychological processes underpinning and informing technological developments as they pertain to this growing, but now well-established industry. In other words, you, the readers, are a diverse bunch. Accordingly, the chapters presented here are also diverse, and it is possible that you will find some more-or-less suited to your own perspectives. The challenge as editors was to create at least one pathway through the material from our authors, and in a sense one might consider your journey through the book analogous to playing a game; there is a narrative with a beginning (a historical overview of past practices), a middle (existing techniques), and an end (in some ways the trickiest part of this book, as in this case, the end is not an end but rather a look towards further developments). The future, of course, requires us to use our 'crystal ball' to some extent, and provide examples which we feel are well grounded in existing work pertaining to emotion as they might then be adapted to video game soundtracking.

What Is the Field?

If you are reading this book, you will likely need no introduction to the world of video games. Video game soundtracking, and particularly the study of emotion in relation to the practice, does require some clarification, however. We might begin by stating what this field is *not*. There is a growing field of interdisciplinary study, described by its proponents as ludomusicology (van Elferen 2014; Summers 2016), which is concerned with a very specific approach to studying video game music. This approach borrows from traditional musicological analysis and is specifically concerned with the music involved in video games, whereas our approach is to view the complete soundtrack (i.e., both music, speech and sound effects), through the lens of emotion. Studies of music and emotion abound in psychological literature, and of course music is often described as 'the language of emotion'. There can be little doubt that music can be used to augment and enhance the viewers' emotional responses in film narrative, and that done well, this effect can be significant in conjunction with gameplay. Our approach also considers the act of production: developing tools for interacting, metering emotional responses, and indeed producing emotion-driven soundtrack elements or complete scores. Various chapters address different stages of the production chain as it pertains to emotion in videogame soundtracking. What we tend to stay away from—by design—is the musicological post-analysis type of approach that you might find in ludomusicology. For such work, the interested reader might be well directed to recent publications such as *Ludomusicology: Approaches to Video Game Music* (Summers 2016) which has emerged from the academic community involved in ludomusicology as a formal discipline.

Defining Emotion

It is therefore also important to begin this book with a subsequent definition, considering emotion, as the lens through which we will examine the various chapters which follow. Again, we borrow heavily here from psychology and cognitive sciences, as one must when addressing interdisciplinary work which remains somewhat in its infancy. Whilst Chap. 2 will give a fuller treatment to differing views of emotion in music from the literature, we will here give a short definition to begin proceedings, on the proviso that the reader need not fully agree with this definition, and that it may remain somewhat in flux as you progress through the narrative of the book.

Overview of this Book

This book is divided chronologically into three sections, past, present, and future. The specifics of each section are hopefully self-explanatory, but the reader should note that as with every emerging field, the dividing lines between these categories can sometimes be blurry. We would like to take a moment here to consider some of the specific chapters, and explain the thinking behind the overall narrative of the book. Recent scholarship, in the form of ludomusicology, has begun to achieve something akin to a critical mass. At the same time, a challenge, for such scholarship and theory, is the rapid pace with which relevant technologies change and develop (e.g. a significant and growing proportion of gaming activity is now undertaken via smart-phones and tablets). A further example is afforded by the release of commercial virtual reality gaming. It will therefore be no surprise to see chapters which address spatial sound, particularly 3-dimensional sound, as in the chapter by Damian Murphy and Joseph Rees-Jones, who evaluate methods for positioning sound objects in space in videogames, virtual reality systems and virtual auditory environments. Their chapter looks closely at the practicality and trade-offs of using these methods in terms of perceived source locatedness and support for moving sound sources and listeners in virtual environments. They also offer an exciting overview of the potential of gameplay advantages through the positioning, spread and movement of the sound source, creating a new paradigm for analyzing audio in gameplay. In the future, virtual reality environments and the use of machine intelligence will help to create embodied interactions in the player—surely one of the most exciting and immersive experiences a player might engage with is the feeling of truly being embodied in the gameplay, and we consider this from a theoretical perspective in Chap. 5. The theme of motion and embodiment is continued by Federico Visi's chapter, which focuses on the measurement of motion, using motion capture to control procedural audio and real-time signal processing synthesis, and in particular on extracting motion features from various devices and using them as an input to generative algorithms.

David Bessell, one of the pioneering writers on the subject of video game soundtracking, and an active video game composer, will present a chapter examining a real-world example of composing for a horror game. His chapter explores how the video game dramaturgy, in relation to non-linear musical form, impact on the ability to generate appropriate emotions at appropriate points in gameplay, particularly relevant to horror soundtracking which relies heavily on music to augment emotional impact. Also from the real-world of game-design, Alexis Kirke's chapter explores the world of 'audio-only' games. There has been an increasing popularity in audio-only games, again partly and particularly due to the omniscient access from and to mobile devices. These are usually based on spoken-word and sound-effect scene setting. This chapter investigates whether this can be taken further into games which are entirely abstract sonic and music experiences, and presents a first example based on Minecraft: Musicraft.

Other chapters will address the emerging field of brain-computer interfacing, for both emotion measurement and control signal generation, the use of algorithmic composition techniques for soundtrack generation in the context of MMORPG, a historical overview of emotion-driven soundtrack generation strategies (and shortcomings), and a glance to the future in terms of using biophysiological interfacing and affectively driven signal processing to generate individually responsive emotion-driven soundtracks. Together, we hope that you will be able to draw inspiration for real-world work, as well as theoretical analysis and frameworking from the more general fields of emotion, biosensing, and neurophysiology, to look towards exciting new pastures for video game soundtracking. Thank you for taking the time to pick up this book.

Sincerely,

Duncan Williams & Newton Lee, Editors

References

van Elferen, I.: Ludomusicology and the new drastic (2014)
On "Videogame" Versus "Video Game".: WIRED. https://www.wired.com/2007/11/on-video-game-ve/ (2017). Accessed 13 July
Summers, T.: Analyzing Video Game Music. Equinox, Sheffield (2016)

Chapter 2
An Overview of Emotion as a Parameter in Music, Definitions, and Historical Approaches

Duncan Williams

Introduction

This chapter presents a theoretical overview of emotion in the context of music, particularly emotional analysis, different types of model, and the distinction between perceived and induced emotions. All are necessary to understand in order to examine emotion in the video game soundtracking context. You may be a videogame designer, sound designer, composer, or player; professional or enthusiastic amateur. Regardless, you will be familiar with the powerful role which soundtracking can play in shaping your experience. Music is a well-documented way to communicate feelings and emotional states, regardless of whether one has written, performed, or simply listened to it. When combined with other modalities, i.e., listening and seeing, or in the case of many games, listening, seeing, and responding with gameplay actions, the experience can become even more intense. High quality soundtracking has the potential to enhance player experience in video games (Grimshaw et al. 2008). Combining emotionally congruent sound-tracking with game narrative has the potential to create significantly stronger affective responses than either stimulus alone—the power of multimodal stimuli on affective response has been shown both anecdotally and scientifically (Camurri et al. 2005). Video game soundtracking has an inexorable link with the available technology at the time of development. This meant that there were—at least—some limitations in terms of what might be achievable in the soundtracking efforts for earlier generations of game, whether that be restrictions based on the type of synthesizer available to the composer, or the storage medium in terms of digital sound effects and speech. Game audio requires at least two additional challenges over other sound-for-picture work; firstly, the need to be dynamic (responding to gameplay states) and secondly to be emotionally

D. Williams (✉)
Digital Creativity Labs, University of York, York, UK
e-mail: duncan.williams@york.ac.uk

D. Williams, N. Lee (eds.), *Emotion in Video Game Soundtracking*, International Series on Computer Entertainment and Media Technology,
https://doi.org/10.1007/978-3-319-72272-6_2

congruent whilst adapting to non-linear narrative changes (Collins 2008). Early solutions such as looping can be come repetitive, and ultimately break player immersion, but branching strategies (where different cues are multiplexed at narrative breakpoints) can drastically increase the compositional complexity required in the music implementation when creating a soundtrack (Lipscomb and Zehnder 2004).

This chapter explores definitions for these terms, and a few traditional approaches to achieving such goals. Of course, for practical reasons, it would be impossible to include every example of novel game soundtracking with regards to emotional content, therefore this chapter will consider one as a starting point from which the interested reader may continue to explore. Our example is Lucasarts implementation of a dynamic system, *iMuse* (see Strank (2013), for a full treatment) to accompany their role-playing games series in the late 1980s (which included the *Indiana Jones* series and perhaps most famously, the *Monkey Island* series of games) (Warren 2003), which this chapter will explore in some more detail. This system implemented two now commonplace solutions, horizontal re-sequencing and vertical re-orchestration, both of which were readily implementable due to the use of MIDI orchestration (Huron 1997). However, the move towards real audio made many of these transformations more complex beyond the compositional aspect alone. This chapter considers how a combination of factors, not least historical improvements in storage space and audio quality both address and create such difficulties.

Music and Emotion

This chapter Music has been shown to induce physical responses on a conscious, and unconscious level (Grewe et al. 2005; Grewe et al. 2007). Such measurements can be used as indicators of affective states. Emotional assessment in empirical work often makes use of recorded music (Wedin 1969, 1972; Gabrielsson and Lindström 2001; Gabrielsson and Juslin 2003) or synthesised test tones (Scherer 1972; Juslin 1997) to populate stimulus sets for subsequent emotional evaluation. There are some reported difficulties with such evaluations with specific regards to measurement of emotional responses to music. For example, some research has found that the same piece of music can elicit different responses at different times in the same listener (Juslin and Sloboda 2010). If we consider a listener who is already in a sad or depressed state, it is quite possible that listening to 'sad' music may in fact increase listener valence. Another challenge for such evaluations is that music may intentionally be written, conducted, or performed in such a manner as to be intentionally ambiguous. Indeed, perceptual ambiguity might be considered beneficial as listeners can be left to craft their own discrete responses (Cross 2005). The breadth of analysis given to song lyrics gives many such examples of the pleasure listeners can take in deriving their own meaning from seemingly ambiguous music

Before examining existing systems and considering the future of video game soundtracking, we must first define the terminology that will be used, including the various psychological approaches to documenting musical affect, and the musical and acoustic features that such systems utilize.

Types of Emotion

Literature concerning the psychological approaches to musical stimuli broadly documents three types of emotional response, each increasing in duration:

1. *Emotion*: A short-lived episode; usually evoked by an identifiable stimulus event that can further influence or direct perception and action. It is important to note here although listeners may experience emotions in isolation, the affective response to music is more likely to reflect collections and blends of emotions.
2. *Affect*/Subjective feeling: Longer than an emotion, affect is the experience of emotions or feelings evoked by music in the listener.
3. *Mood*: Longer-lived again, mood is a more diffuse affective construct; usually latent and indiscriminate as to eliciting events; mood may influence or direct cognitive tasks.

The distinction between emotion and subjective feeling (the part of emotion which is consciously accessible to the person experiencing the emotion), whereas the other types of affective response are not necessarily available to conscious report and have practical consequences over components of emotion (such as motor expression or action tendencies). In any case, measuring these responses is difficult—the former components may or may not be consciously perceived, or correctly reported by the person experiencing the emotion. Bodily symptoms alone are not sufficient to evoke and consequently allow for the reporting of emotions (Schachter and Singer 1962). A full treatment of the underlying mechanisms at play in accounting for evaluation of musical emotion is given by Juslin and Västfjäll (Juslin and Västfjäll 2008).

Models of Emotion

There are two main types of emotional models used in relation to affective analysis of music—categorical and dimensional models. Categorical models use discrete labels to describe affective responses. Dimensional approaches attempt to model affective phenomenon as a set of coordinates in a low-dimensional space (Eerola and Vuoskoski 2010). Discrete labels from categorical approaches (for example, mood tags in music databases) can often be mapped onto dimensional models, giving a degree of convergence between the two. Neither are music-specific emotional models, but both have been applied to music in many studies. More recently, music-specific approaches have been developed (Zentner et al. 2000, 2008).

The circumplex dimensional model, for instance, describes the semantic space of emotion within two orthogonal dimensions, valence and arousal, in four quadrants (e.g. positive valence, high arousal). This space has been proposed to represent the blend of interacting neurophysiological systems dedicated to the processing of valence (pleasure–displeasure) and arousal (quiet–activated) (Russell 1980, 2003). The Geneva Emotion Music Scale (GEMS) describes nine dimensions that represent the semantic space of musically evoked emotions (Scherer 2004), but unlike the circumplex model, no assumption is made as to the neural circuitry underlying these semantic dimensions. GEMS is a measurement tool to guide researchers who wish to probe the emotion felt by the listener as they are experiencing it. The same researchers devised experiments which examined the differences between *felt* and *perceived* emotions.

> Generally speaking, emotions were less frequently felt in response to music than they were perceived as expressive properties of the music. (Zentner et al. 2008, p. 502)

This distinction has been well documented, for example see (Scherer 2004; Marin and Bhattacharya 2010; Daly et al. 2015) though the precise terminology used to differentiate the two varies widely. Perhaps unsurprisingly, results tying musical parameters to induced or experienced emotions do not often provide a clear description of the mechanisms at play, and the terminology used is inconsistent. n induced emotion would be an affective state experienced by the listener, rather than an affect which the listener understands from the composition—by way of example, this would be the difference between listeners reporting that they have 'heard sad music' rather than actually 'felt sad' as a result of listening to the same.

For a more complete investigation of the differences in methodological and epistemological approaches to perceived and induced emotional responses to music, the reader is referred to Scherer (2004).

Sad Music

A growing body of research suggests that empathy is one of the key factors which dictates the success of emotional communication from music (Bråten 2007; Arizmendi 2011). The voice has been the subject of significant acoustic analysis in order to determine prosodic cues (Barkat et al. 1999; Bach et al. 2008) and findings suggest some level of universal understanding of these cues. Intuitively, for example, an angry voice will be louder, perhaps brighter (a higher spectral centroid) and with a faster rate of temporal cues. However, beyond such acoustic cues and their affective connotations in the prosody of the voice, singing also contains another emotional signpost: lyrics. Often it is enough to simply read lyrics to determine their emotional quality, but when lyrics are matched with congruent acoustic or musical cues in the vocal delivery, for example, happy lyrics accompanied by a major key as in Bernstein and Sondheim's "I Feel Pretty" from the musical West Side Story (1957), an unambiguous affective reading is readily identifiable:

I feel pretty,
Oh, so pretty,
I feel pretty and witty and bright!
And I pity
Any girl who isn't me tonight.

However, the affective content of a lyric can also be contradicted by the musical features used in its delivery. An example of this can be heard in Lesley Gore's "It's My Party", from I'll Cry If I Want to (1963), which sets lyrics relaying the distress of a heartbroken teenager to an incongruous major key (A Major). The resulting affect becomes unclear to the listener, though the song has maintained an enduring popularity in any case. David Bowie's "Heroes" (1977) illustrates the spread of this kind of affective ambiguity. Despite the dark subject matter relating to alcoholism and the breakdown of a relationship, it is routinely used as a soundtrack for celebrations, major sporting events, and the like. The vocal recording process, led by producer Tony Visconti, made use of a novel system of processing whereby a number of discrete microphones were used to capture Bowie's vocal, each positioned further from the singer. These ambient microphones were then muted during the quieter, opening passages of the vocal, and gradually introduced as the vocal became more intense—multitracking was not used to achieve this effect, as the recording sessions were subject to the restrictions of analogue technology at the time. The sonic parallel between the characters in Bowie's lyrics and the increased ambient sound in the final sections of the vocal, which, thanks to Visconti's microphone technique becoming continually more lost and set back in the wall-of-sound production, is quite striking.

The question as to why a listener might deliberately choose to listen to singing which reflects a negative state—in other words, explicitly sad music—has recently been the subject of much investigation by music psychologists (Vuoskoski and Eerola 2012). The popularity of sad music again suggests that empathy, particularly in the voice, can be a powerful trigger for the listener. Leonard Cohen's "Famous Blue Raincoat", from Songs of Love and Hate (1971) makes use of a number of affective correlates for negative valence, performed as it is in A minor, with a slow tempo and a ¾ time signature, the lyrics are mainly accented as amphibrachs (˘ ¯ ˘) over the meter with limited melodic leap and a lilting, low volume. A full analysis of the song and Cohen's use of the link between lyrics and structure to create affective meaning is given by Christophe Herold in (Herold, n.d.). Again, the performance has endured a lasting popularity.

The MIR (music information retrieval) community have made significant inroads into affective labelling of music databases, in order to allow listeners to select music according to a range of moods, of which sad is often. The utility of such applications suggests that we might conclude that a listener in a negative state of mind may enjoy listening to music that reflects the same affective state in the same way that a conversation with an empathetic listener might be pleasurable, or that mirroring can generate rapport in broader communication analyses.

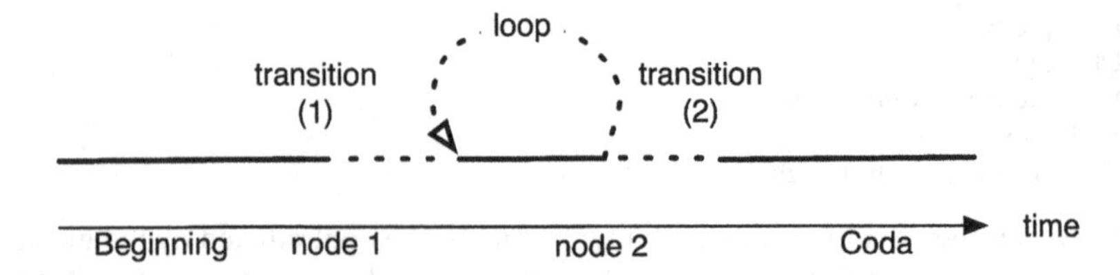

Fig. 2.1 A game soundtrack adapting to play dynamically—the middle section is looped until a narrative breakpoint is reached. All passages, and transitions between passages (cadences etc) are precomposed. This approach has been used extensively in a wide range of games (see, for example, the *Final Fantasy* series)

Immersion vs Emotional Congruence

Creating congruent sound-tracking for video games is a non-trivial task, as their interactive nature necessitates dynamic and potentially non-linear sound-tracking. This requirement is essentially due to the unpredictable element of player control over the narrative, without which the game would cease to be interactive. This problem has been approached with various solutions. A commonly used solution, as shown in Fig. 2.1, is to loop a pre-composed passage of music until a narrative break, such as the end of a level, death of a player, victory in a battle, and so on. However, this approach can become repetitive, and potentially irritating to the player if the transition points are not carefully managed, as musical repetition has been shown to have its own impact on the emotional state of the listener (Livingstone et al. 2010).

The relative importance to the gamer of immersion and emotional congruence is not necessarily evenly weighted. Immersion is essential, or the player will likely cease playing the game; keeping the player hooked is an important goal of game design. Emotional congruence, on the other hand, may enhance player immersion, but is likely to hold a lower place in the player's perceptual hierarchy. One notable exception might be in situations where the player deliberately controls the music (as can be seen in games like *Guitar Hero*, for example). The process then becomes a complex feedback loop wherein the player not only influences the selection of music according to their mood but the selection of music also has a subsequent impact on the player's mood. Considering the *affective potential* of music in video games is a useful way of understanding, and potentially enhancing, the player experience of emotion from the gameplay narrative. This term will be fully explored in Chap. 3.

An alternative solution would be to use divergent musical sequences to create a less repetitive soundtrack, as shown in Fig. 2.2. However, the trade-off becomes one of practicality and complexity, as each divergent branch requires both storage space on the physical medium (which is often at a premium in the world of video gaming), and a human cost in the time taken to compose enough material.

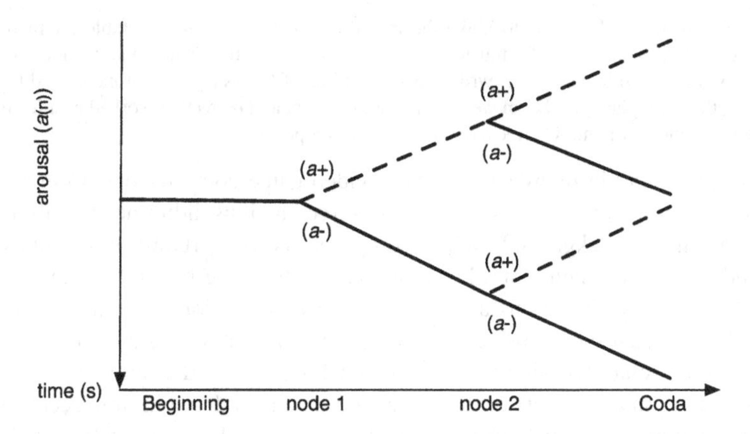

Fig. 2.2 Divergent branching system used to sequence a musical score without looping, determined by arousal cues as meta-tags corresponding to gameplay narrative (arousal increase or decrease at beginning, middle, and end). Even for a very simple narrative with just three stages, 7 discrete pieces of music are required, with a resulting impact on composition time and storage space

An Example: iMuse and *Monkey Island*

iMuse, the music generation system accompanying SCUMM, or the *script creation utility for maniac mansion* (*Maniac Mansion* was a text-driven adventure game commissioned by Lucasarts in the 1980s), facilitating a specific character and narrative driven music performance which continued throughout the gameplay, which allowed for on-the-fly generation of emotionally congruent music, within specific rulesets. *The Secret of Monkey Island* was the first title to really explore this potential, and has enjoyed a long-lived popularity, including being one of only five games featured at the Art of Video Games exhibition at the Smithsonian in 2012 (alongside *Pacman*, *Super Mario World*, *Myst*, and *Flower*). Clearly, this puts this adventure title in some illustrious company. The iMuse engine is now available as a Java VirtualMachine for readers who are eager to experiment with the ruleset themselves.

Composer Michael Land was determined to synchronize gameplay action with emotionally congruent music. His soundtrack cues used both characterization (specific tunes for specific characters and venues) as well as localization cues to generate his responsive soundtrack on-the-fly. Originally rendered in MIDI, this strategy required a large number of complex branches when it was re-implemented with 'real' audio (i.e., human musicians performing each branch which would then be recorded as audio files, and cross-faded between depending on the gameplay cues). The game series has recently been rebooted on both iPhone and console platforms, as well as OSx and iPad implementations. Other games which use the same technology from the stable include *Grim Fandango* (PS4 2015), and perhaps unsurprisingly given the intellectual property owners base, *Star Wars: Tie Fighter, Full Throttle* etc.

> The goal behind iMuse was to make the music respond to the unpredictable interactive changes in the game as if the music had been composed with advance knowledge about what was going to happen… … write music with lots of layers and lots of paths, and then during the gameplay the system would choose the layers and paths that worked best for the situation. Michael Land, Composer and engine developer

In the past (and in many cases today), video game compositions differed from that of television and movies in that the on-screen actions didn't necessarily match the music in the background. While composers could certainly write tunes that matched the general theme of a level, they didn't have the foresight or technology to predict the player's actions and change their compositions accordingly. Theme and variation was used extensively by Land to extend characterization and narrative options from relatively simple compositional ideas. Land's strategies included dynamic solutions: elements of the score would be re-ordered in time according to a player action or specific narrative, and re-sequenced vertically (i.e., different instruments added, timbres, or volume selections applied—strategies which have now become commonplace in the use of *procedural audio* for video game soundtracking (Plans and Morelli 2012). It is easy to see why this might present technical challenges: for example, when recording musicians there must be a significant amount of acoustic isolation in order to achieve successful vertical re-orchestration on-the-fly. However, such techniques also present compositional challenges. For example, if one is to create a situation where the score must be re-sequenced in either time or instrumentation, the basic melody and harmony cannot feature modulation or changes in tempo. Why is this a problem? Both are amongst the biggest emotional cues for listeners and players (Dickinson 2008; Beveridge and Knox 2010). Listener expectations extend beyond tempo and mode, and often the emotional effect of specific cadences and endings is significant—and again, requires specific preparation in the design and development of a video game soundtrack to circumnavigate (Berg and Wingstedt 2005; Huron 2006). Cadences which are abruptly caused by gameplay state changes will jar the play and cause immersion to be reduced. An undesirable attribute in a soundtrack, to say the least. The next chapter will explore implementations of these concepts in current and future work.

References

Arizmendi, T.G.: Linking mechanisms: emotional contagion, empathy, and imagery. Psychoanal. Psychol. **28**, 405 (2011)

Bach, D.R., Grandjean, D., Sander, D., Herdener, M., Strik, W.K., Seifritz, E.: The effect of appraisal level on processing of emotional prosody in meaningless speech. NeuroImage. **42**, 919–927 (2008)

Barkat, M., Ohala, J., Pellegrino, F.: Prosody as a distinctive feature for the discrimination of arabic dialects. In: EUROSPEECH '99, pp. 395–398. (1999)

Berg, J., Wingstedt, J.: Relations between selected musical parameters and expressed emotions: extending the potential of computer entertainment. In: Proceedings of the 2005 ACM SIGCHI International Conference on Advances in Computer Entertainment Technology, pp. 164–171. ACM, Valencia, Spain (2005)

Beveridge, S., Knox, D.: Emotion classification of western contemporary music: identifying a representative feature set. In: Proceedings of the 11th International Conference on Music Perception and Cognition, Seattle, Washington, 23–27 August 2010
Bråten, S.: On Being Moved: From Mirror Neurons to Empathy, vol. 68. John Benjamins, Philadelphia (2007)
Camurri, A., Volpe, G., De Poli, G., Leman, M.: Communicating expressiveness and affect in multimodal interactive systems. IEEE Multimedia. **12**(1), 43–53 (2005)
Collins, K.: Game Sound: An Introduction to the History, Theory, and Practice of Video Game Music and Sound Design. MIT Press, Cambridge, MA (2008)
Cross, I.: Music and meaning, ambiguity and evolution. In: Musical Communication, pp. 27–43. Oxford University Press, New York (2005)
Daly, I., Williams, D., Hallowell, J., Hwang, F., Kirke, A., Malik, A., Weaver, J., Miranda, E., Nasuto, S.J.: Music-induced emotions can be predicted from a combination of brain activity and acoustic features. Brain Cogn. **101**, 1–11 (2015). https://doi.org/10.1016/j.bandc.2015.08.003
Dickinson, K.: Off Key: When Film and Music Won't Work Together. Oxford University Press, New York (2008)
Eerola, T., Vuoskoski, J.K.: A comparison of the discrete and dimensional models of emotion in music. Psychol. Music. **39**, 18–49 (2010). https://doi.org/10.1177/0305735610362821
Gabrielsson, A., Juslin, P.N.: Emotional expression in music. In: Davidson, R.J., Scherer, K.R., Goldsmith, H.H. (eds.) Handbook of Affective Sciences Series in Affective Science, pp. 503–534. Oxford University Press, New York (2003)
Gabrielsson, A., Lindström, E.: The influence of musical structure on emotional expression. In: Juslin, P.N., Sloboda, J.A. (eds.) Music and Emotion: Theory and Research Series in Affective Science, pp. 223–248. Oxford University Press, New York (2001)
Grewe, O., Nagel, F., Kopiez, R., Altenmüller, E.: How does music arouse "chills"? Ann. N. Y. Acad. Sci. **1060**, 446–449 (2005)
Grewe, O., Nagel, F., Kopiez, R., Altenmüller, E.: Emotions over time: synchronicity and development of subjective, physiological, and facial affective reactions to music. Emotion. **7**, 774–788 (2007). https://doi.org/10.1037/1528-3542.7.4.774
Grimshaw, M., Lindley, C.A., Nacke, L.: Sound and immersion in the first-person shooter: mixed measurement of the player's sonic experience. In: Proceedings of Audio Mostly Conference, 1–7. Lulea University of Technology, Piteå (2008)
Herold, C.: Famous Blue Raincoat:-an approach to an integrated analysis of a song (n.d.)
Huron, D.: Humdrum and Kern: Selective Feature Encoding, Beyond MIDI: The Handbook Of Musical Codes. MIT Press, Cambridge, MA (1997)
Huron, D.B.: Sweet Anticipation: Music and the Psychology of Expectation. MIT Press, Cambridge, MA (2006)
Juslin, P.N.: Emotional communication in music performance: a functionalist perspective and some data. Music. Percept. **14**, 383–418 (1997)
Juslin, P.N., Sloboda, J.A.: Handbook of Music and emotion: Theory, Research, Applications. Oxford University Press, Oxford (2010)
Juslin, P.N., Västfjäll, D.: Emotional responses to music: the need to consider underlying mechanisms. Behav. Brain Sci. **31**, 559 (2008). https://doi.org/10.1017/S0140525X08005293
Lipscomb, S.D., Zehnder, S.M.: Immersion in the virtual environment: the effect of a musical score on the video gaming experience. J. Physiol. Anthropol. Appl. Hum. Sci. **23**(6), 337–343 (2004)
Livingstone, S.R., Muhlberger, R., Brown, A.R., Thompson, W.F.: Changing musical emotion: a computational rule system for modifying score and performance. Comput. Music. J. **34**(1), 41–64 (2010)
Marin, M.M., Bhattacharya, J.: Music Induced Emotions: Some Current Issues and Cross-Modal Comparisons. Nova Science, New York (2010)

Plans, D., Morelli, D.: Experience-driven procedural music generation for games. IEEE Trans. Comput. Intell. AI Games. **4**(3), 192–198 (2012). https://doi.org/10.1109/TCIAIG.2012.2212899

Russell, J.A.: A circumplex model of affect. J. Pers. Soc. Psychol. **39**, 1161 (1980)

Russell, J.A.: Core affect and the psychological construction of emotion. Psychol. Rev. **110**, 145 (2003)

Schachter, S., Singer, J.: Cognitive, social, and physiological determinants of emotional state. Psychol. Rev. **69**, 379–399 (1962). https://doi.org/10.1037/h0046234

Scherer, K.R.: Acoustic concomitants of emotional dimensions: Judging affect from synthesized tone sequences. In: Proceedings of the Eastern Psychological Association Meeting. Education Resources Information Center, Boson (1972)

Scherer, K.R.: Which emotions can be induced by music? What are the underlying mechanisms? And how can we measure them? J. New Music Res. **33**, 239–251 (2004)

Strank, W.: The legacy of IMuse: interactive video game music in the 1990s. In: Moormann, P. (ed.) Music and Game, pp. 81–91. Springer VS, Wiesbaden (2013)

Vuoskoski, J.K., Eerola, T.: Can sad music really make you sad? Indirect measures of affective states induced by music and autobiographical memories. Psychol. Aesthet. Creat. Arts. **6**, 204 (2012)

Warren, C.: LucasArts and the design of successful adventure games: the true secret of Monkey Island. http://www-sul.stanford.edu/depts/hasrg/histsci/STS145papers/Warren.pdf (2003)

Wedin, L.: Dimension analysis of emotional expression in music. Swed. J. Musicol. **51**, 119–140 (1969)

Wedin, L.: A multidimensional study of perceptual-emotional qualities in music. Scand. J. Psychol. **13**, 241–257 (1972). https://doi.org/10.1111/j.1467-9450.1972.tb00072.x

Zentner, M., Grandjean, D., Scherer, K.R.: Emotions evoked by the sound of music: characterization, classification, and measurement. Emotion. **8**, 494–521 (2008). https://doi.org/10.1037/1528-3542.8.4.494

Zentner, M.R., Meylan, S., Scherer, K.R.: Exploring musical emotions across five genres of music. In: Sixth International Conference of the Society for Music Perception and Cognition (ICMPC), 5–10 Aug 2000

Chapter 3
Emotion in Speech, Singing, and Sound Effects

Duncan Williams

Introduction

This chapter explores the theoretical context of emotion studies in terms of speech and sound effects, and in particular the concept of *affective potential*. Voice actors in game soundtracking can have a particularly powerful impact on the emotional presentation of a narrative; and this affective control can go beyond that of the actor alone if combined with emotionally-targeted signal processing (for example, sound design and audio processing techniques). The prospect of synchronousing emotionally congruent sound effects remains a fertile area for further work, but an initial study which will be presented later in this chapter suggests that timbral features from speech and sound effects can exert an influence on the perceived emotional response of a listener in the context of dynamic soundtracking for video games. This chapter extends upon material originally presented at the Audio Engineering Society conference on video game soundtracking in London, UK, 2015 (Williams et al. 2015), and subsequently on the specific design of affect in vocal production at the Audio Engineering society convention in New York, 2015 (Williams 2015a). Prosodic (nonverbal) speech features have been the subject of a considerable amount of research (Gobl 2003; Pell 2006). The role of such features as a communicative tool in emotion studies suggests that acoustic manipulation of prosody could be a useful way to explore emotional communication (Frick 1985; Baum and Nowicki 1998). For example, in studies which use dimensional approaches to emotion, known acoustic correlations found in prosody include emotional arousal with pitch height, range, rate of speech, and loudness. Some emotional cues can be derived acoustically from prosody (Bach et al. 2008) by time series analysis in a manner which is analogous to the temporal characteristics used to determine such cues in

D. Williams (✉)
Digital Creativity Labs, University of York, York, UK
e-mail: duncan.williams@york.ac.uk

D. Williams, N. Lee (eds.), *Emotion in Video Game Soundtracking*, International Series on Computer Entertainment and Media Technology,
https://doi.org/10.1007/978-3-319-72272-6_3

musical sequences (Gobl 2003; Juslin and Laukka 2006; Kotlyar and Morozov 1976; Deng and Leung 2013), for example pitch height and range, loudness, and density are suggested to correlate strongly with affective arousal by some research.

Affective Potential, and the Power of the Voice

Affective science increasingly concludes that the voice is a powerful tool for emotional communication. The process of creating a video game voice through studio recording gives listeners the opportunity to engage in experiences of character voices which are quite unlike that which would be achieved in a traditional interaction when compared with day-to-day speech. The audio production chain, from sound capture using particular sound recording techniques, to specific effects processing affords the engineer (or the voice artist themselves) unparalleled access to shape the end product, and thereby enhance the affective impact of the voice for the listener.

The voice itself, central to recorded music of many genres, can be processed by technology in an increasing number of ways, allowing for deliberate shaping of emotional character, and perhaps, ultimately, for stronger emotional communication with the listener, or in most cases found in this book, the gamer themselves. We consider this *affective potential*, as a way of considering how the sound designer might deliberately manipulate the voice to communicate emotional intent (Garrard and Williams 2013). The effect of these adjustments on perceived affect is both somewhat subjective, and open to question, but the hope of this chapter to provide a starting point for marrying technical decisions with philosophical outcomes in the context of affectively-driven analysis of soundtracking in videogames out with music; that found in speech and sound effects.

Considering parameterisation of the affective potential of the voice is a hypothetically useful way of facilitating a sound designers practice by means of understanding a gamers experience of emotion from the finished product, the game soundtrack. For the purposes of this chapter, the term ‘affective attribute’ should be taken to mean a verbal descriptor which existing researchers have shown contributes in whole or in part to the listener’s perception of affect. The term ‘affective correlate’ should be taken as any acoustic property, or combination of properties as is the case of many musical features, that is shown to correspond to the listener’s perception of a specific affective attribute, group of attributes, or affective response as a whole (Le Groux and Verschure 2010; Coutinho and Cangelosi 2011; Daly et al. 2015). Some correlates might be considered musical or acoustic, such as brightness, a timbral descriptor which in turn has previously been shown to have a strong correlation to a single acoustic property, spectral centroid (Brookes and Williams 2007). Many of these correlates might be controlled by the voice to deliberately target affective responses, and some can be further manipulated by technology in the recording process for example, dynamic range manipulation, reverberant environment (Mo et al. 2016). Thus, in the terminology adopted here, reverberation size and balance could be a direct affective correlate for several attributes.

The voice actor themselves can introduce a significant amount of affective variation through their own delivery (louder, quieter, different timbres and so on), other factors can also influence the affective potential of the recording, not least the lyrics and their relationship to the music, but also the way in which technology might be used to capture the performance (from microphone placement, to overdubbing and multitracking techniques). The performance element of a vocal production can involve much more than simply what happens with the physical apparatus (glottis, vocal chords, mouth and so forth), and indeed the physical apparatus is an area which is not directly influenced by technology (at least in typical music making). More likely variation is found in the recording environment in terms of space and frequency response, and microphone selection and placement, both of which offer a large range of possibilities to the voice recording, and the capability to change the way the voice subsequently communicates to the listener.

The power of the voice to mimic other sounds and create extraordinary sonic worlds through call and response has been well demonstrated by Trevor Wishart's voice morphing pieces (Wishart 1988). Wishart makes use of careful positioning of a standard dynamic microphone, in close proximity to the mouth but facing down so as to minimize unwanted plosives. The proximity effect, whereby low frequencies tend to appear to be amplified when a microphone is moved closer to the sound source (Josephson 1999) can give the voice acoustic properties which it did not possess from a distance, but moreover the use of close microphone techniques on the voice gives a unique listening impression which would never occur in day-to-day life (regular conversations never involve positioning oneself directly in front of the mouth of the other speaker, for example), and thus the listening to the timbres of the recorded voice becomes a wholly different experience for the listener, helping the voice to communicate emotions with a great degree of intensity.

The practice of multitracking vocals gives a way to create new timbres and effects which could not be achieved with a single track, and as in the case of flanging informed the development of new technological tools. However, layering different vocal registers and deliveries offers an entirely different result. This technique can also facilitate more creative vocal production when it is used to layer different vocal timbres, such as spoken, whispered, and falsetto lines. No particular special effects processing, other than the power of the multitrack recorder to layer these timbres is used, but the sound world created by the addition of whispering and spoken timbres is immediately novel compared with what might be achieved by means of a single, solo performance—and, by the same token, the realism of the performance is also compromised. This is not to say that realism and multitracked performance must be mutually exclusive. However, there are a significant number of well-documented challenges involved in capturing realism whilst multitracking, including many that are specifically related to the consistent and congruent communication of emotion by ensembles.

The goal of capturing a natural performance has been explored as both desirable or, contrastingly, unnecessarily limiting in the context of vocal affect, Producers have a number of methods for dealing with this, including the use of guide tracks where the whole ensemble can perform together and subsequently replace some or all of the performances in question, but the psychological techniques required by both producer and performer in order to achieve performances which are emotionally congruent between vocalist and the remainder of the ensemble in a multitrack situation are manifold and non-trivial. On the other hand, some self-producing musicians have used multitracking to create performances of enormous complexity without the need to have other performers on the 'same emotional page'.

Whilst careful microphone placement, multitracking, and of course the performance of a vocal itself can have a large impact on the affective potential of a recording, advances in recording technology in the past 5 decades have seen an even greater variety of technological processes be deployed in vocal production. Many of these innovations were in fact the product of the analogue recording age, though digital signal processing has undoubtedly given access to these techniques to a much wider range of performing musicians. This section explores this part of the production, the processing and playback stage which occurs after a vocal has been recorded, and considers the impact on affective correlates that processing in the spectral, temporal, and spatial realms, can facilitate.

Compression is a signal processing technique for reducing dynamic range, wherein amplitude peaks are attenuated such that the relative difference between high and low level signals is reduced. Dynamic range compression can occur both before and after the analogue to digital conversion involved in modern recording. If applied before, the effect is often used to optimize recording levels, but the most common application is typically subtler: a reduction in dynamic range variation, and typically an increase in overall perceived loudness as the overall signal is often then applied to further amplification.

The dynamic range of the human auditory system is extremely sensitive, and we can differentiate between a huge variety of signals (often, the roar of a jet aircraft is given as an example of the loud end of our scale, with the sound of a pin dropping at the quiet end). However, the range of comfortable listening is much less than this, and indeed, current professional audio equipment is significantly less capable of dealing with this range of amplitudes, hence the use of dynamic range compression in the mastering stage of audio production, and at some stages of the broadcast chain. However, increasingly, research suggests that dynamic range reduction leads to a longer-term decrease in listener enjoyment.

Vocal production can make use of compression both correctively (in order to compensate for unintended variations in amplitude in the performance), or creatively, whereby the increase in perceived loudness can also have interesting timbral effects such as increasing punch, clarity, or bringing a vocal to the front of a mix in comparison to instrumentation. The trade-off is between these perceptual attributes and realism.

Frequency Response

Selective amplification or attenuation of particular ranges of the frequency spectrum was originally developed in equalizer units as a way of improving telephony, by compensating for uneven frequency responses inherent in passive transmission lines. However, this technology was adopted in popular music recording for both corrective and creative purposes. An unintended by-product of close microphone recording can be temporal distortion on plosive sounds, and sibilance (particularly with female vocals which typically have a higher frequency range than male vocals). These artefacts can be somewhat reduced by careful use of filtering, or by combining filtering with dynamic range control as is the case with de-essing, a process whereby the threshold of a compressor is adjusted according to amplitude in a particular frequency range which corresponds to the vocalists sibilant range. Filtering can be used creatively to imply place and space through selective manipulation of the vocal frequency response.

Spatial Manipulation

Spatial representation in virtual reality environments is addressed more specifically in Chap. 10. Spatial representations derived by reverberation can give specific associations to the listener as perceptual analogies: near, far, big, small, even specific materials can be implied from the reverberant environment. There is a distinction to be drawn between the real spaces which might be used at the performance stage of the vocal production, and the later addition of artificial reverberation to create a virtual sound world around the recorded voice. The latter gives the ability to create imagined, conceptual spaces to accompany the narrative of the voice, and potentially augment it if the spaces are congruent with the affective intention of the vocal content. For example, conveying intimacy with little or no reverberation, or matching a large artificial environment to place the voice further 'back' in the mix might give a remote impression or a large sense of scale. If the space is obviously based on a real-world space (as would be the case with convolution reverberation), it is also likely that the listener might imagine that space, particularly if it is an environment they have some familiarity with (for example, the tight early reflections of singing in a shower cubicle, or the large decay times of a church hall), rather than any particular affective correlation. The influence of spatial manipulation during the recording process itself should not be discounted. Whilst the work of Mo et al. focussed on the affective correlations of reverberation in musical instruments, the placement of reverberation in the monitoring system during a vocal recording is a common practice. The increased confidence which can be instilled in the performer might be related to instrument-derived findings wherein anechoic spaces were felt more 'comic' than spaces with ambience supplied by means of artificial reverberation. However, applying artificial static spaces for the voice to inhabit are not the only means by which spatial impression might be manipulated by technology. Tools which manipulate the specific timing of the voice, to create phasing or flanging effects, using the auditory mechanism to create the illusion of movement, have also been to great effect. Our auditory circuitry makes extensive use of phase correlation between the left and right ear in order to rationalize and localize sounds, and this circuitry can be further manipulated by the use of phase-shifting, a technique where phase interference is used to attenuate or amplify specific frequencies in a signal by summing it with a slightly delayed duplicate. This gives a distinct swirling or swooshing timbre.

Prototype System for Realtime Audio Manipulation in Response to Emotional Trajectory

In order to manipulate perceived emotional responses to specific sounds, acoustic correlates for particular emotional states, or more general target areas on a multidimensional emotion space, must be determined. These could then be used as the target of spectral or temporal acoustic transformation in order to move particular sounds towards narrative-congruent affective intentions. Very recently, series' of acoustic correlates have been examined by means of training regression analysis of time-varying acoustic features with neurophysiological measurements which has revealed some affective correlations in spectral and spectro-temporal acoustic features (Williams 2015b) but in order to be useful in the proposed system, these correlations would need to be extrapolated from testing stimuli with unidimensional variation in a manner which would quantify the acoustic variation (in order to subsequently apply the relevant amount of manipulation to a sound effect to achieve the intended degree of affective change).

In a fully realized affectively-driven sound effect engine, reverberation could reasonably be included as an affective correlate which might be altered on the fly according to particular emotional descriptors as the need arouse in a non-linear soundtrack. However, as with the other possible affective correlates above, this model would need to be the subject of significant further work to in reality; including the search for affective unidimensionality in the reverberant signal properties, and some amount of quantification (e.g., defined by perceptual scaling experiment) across the intended emotional space (or range of emotional descriptors, depending on the particular model of affect being used).

Audio morphing (or sound hybridization, as it is sometimes referred to) is a technique for creating a hybrid sound with characteristics derived from a source and a target sound (Sethares et al. 2009; Olivero et al. 2012). Current systems interpolate between the spectral, or spectral and temporal features of the source and target sound, to create a new feature set which can then steer synthesis. There are various types of audio morphing, including examples where the morph does not take all of the features of the target sound, or where the morph is carried out dynamically and thus particularly applicable to speech processing A parallel can be drawn with the world of visual effects, where one image seamlessly changes shape (morphs) into the other. An opportunity exists to exploit the possibility of harnessing affective correlates as input parameters for a specific type of audio morphing, timbre morphing (Cospito and de Tintis 1998) in order to create on-the-fly emotion driven signal processing—as in the use case under consideration here, this could include sound effect automation in narrative synchronous video game soundtracking as well as other applications where context might change the required emotional content of audio. Experiments such as the one outlined above might help to inform the affective mapping required by such a system. A fully realized system making use of these types of affective quantifications could ultimately work as follows:

1. Analyze the affective state of the current sound file on a 2-D emotion space (as a valence and arousal co-ordinate), and store features via pitch-synchronous FFT (or similar)
2. Input a target affective state co-ordinate (e.g., based on gameplay or other narrative cue)
3. Calculate the difference between the current and target emotional state, using the affective mapping between emotional response and acoustic feature set derived by perceptual experiments
4. Apply spectro-temporal feature morphing to the required degree according to the difference between current state and target (temporal features might make use of dynamic time warping to time-align relevant features before morphing (Caetano and Rodet 2010) spectral features might require subsequent acoustic compensation in the case of unwanted change through overlapping affective correlations)
5. Synthesis affective hybrid sound, e.g., by IFFT (Haken et al. 2007) or other techniques for fast additive synthesis

An overview of this suggested signal flow is shown in Fig. 3.1.

Conclusions

Emotionally congruent sound effect generation can be particularly useful in interactive audio applications with non-linear narrative requirements (such as video gaming). The voice is a powerful tool for emotional communication, especially when paired with affectively-congruent music and sound effects. The process of vocal production comprises, broadly speaking, two stages: performance and processing. Any of these stages might be deliberately manipulated by the game audio designer. A third stage, that of playback, is less often deliberately adjusted by technology to increase the affective response of the listener to the voice, though some changes might be noticed when comparing material in different listening environments, and particularly when listening with headphones.

Acoustic signal processing allows for a level of vocal production which is unparalleled in the natural listening environment, and through particular manipulation of

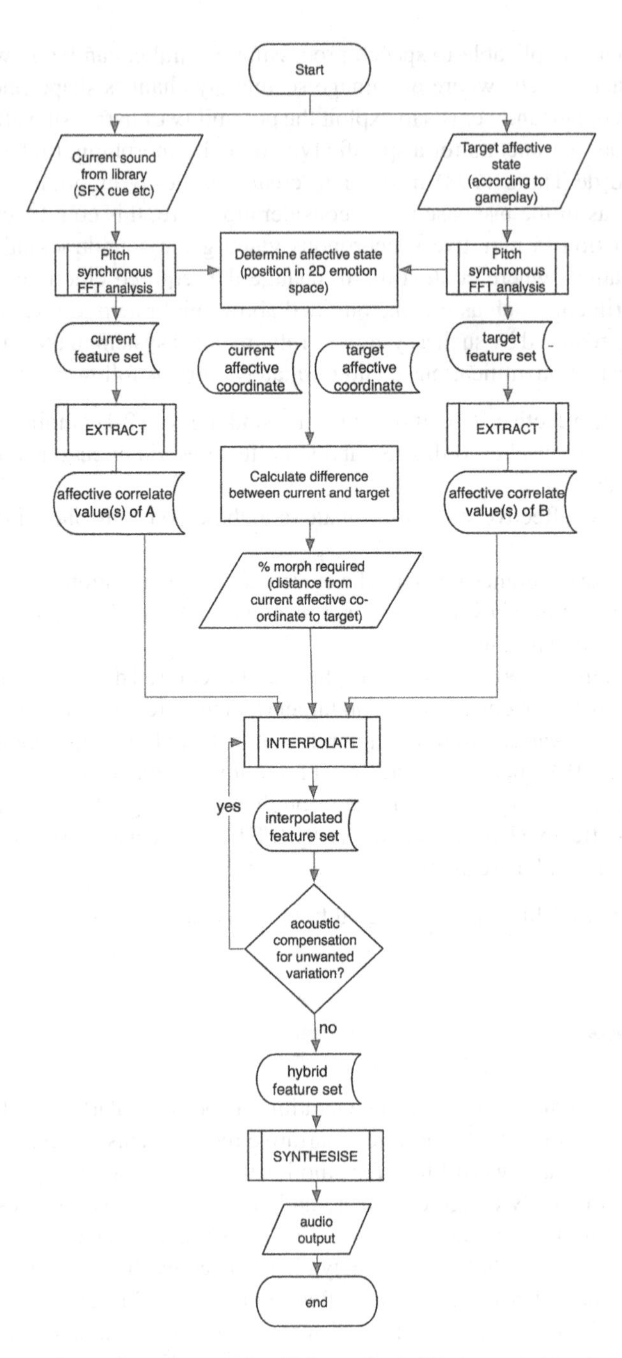

Fig. 3.1 Suggested overview of a system which employs affective mapping of acoustic feature sets to facilitate spectro-temporal morphing between an existing sound library and a gameplay-targetted affective state, for synchronous emotionally congruent soundtrack generation. Specific signal processing routines are capitalized

this processing chain, the affective potential of the recorded voice can be significantly enhanced. These techniques can go beyond corrective processing and be used to craft signature timbres for speech effects. Whilst realism can be compromised by techniques like multitrack recording and layering of vocals, this move away from reality can also be deliberately exploited, such that players have a unique opportunity to engage in a sound world which could not previously be achieved. Affective sciences might find the analysis of such uses of the voice in modern recordings a fertile ground for future investigation of perceived affective correlates in acoustic sources.

Practitioners working in the sound for videogame industry face various practical and aesthetic challenges when creating new sound design for games. Various solutions for synchronous acoustically congruent soundtracking exist, including the use of convolution to manipulate spatial perception in real-time, spectral filtering, Doppler shifting to imply speed and motion and so on. Many of these effects are now available quickly and intuitively (for example in engines like Unreal and Unity). Ascribing and targeting particular emotional states via sound-for-picture is also commonplace (for example, affective soundtracking is regularly used to good effect to heighten the emotional response of an audience in film-making), and has been shown to heighten immersion and engagement in video-game play, but the use of emotional descriptors as control signals for soundtrack manipulation or targetted generation has not yet become commonplace, much less been implemented in the current major audio engines. Previous research has included investigating methods for dynamically generated affectively-charged music in order to target specific emotions congruently and synchronously with the narrative of the gameplay. Such solutions are not trivial because the nature of gaming is that the narrative, and thus the associated soundtracking, will typically change non-linearly under the control of the player. This challenge is also seen at a micro level when considering emotionally congruent, synchronous sound effect generation as a dynamic feature of the overall soundtrack of a game, and thus a system capable of targeting specific emotional responses via sound effect manipulation has obvious advantages in the context of accompanying gaming narrative and enhancing player immersion.

References

Bach, D.R., Grandjean, D., Sander, D., Herdener, M., Strik, W.K., Seifritz, E.: The effect of appraisal level on processing of emotional prosody in meaningless speech. NeuroImage. **42**, 919–927 (2008)

Baum, K.M., Nowicki Jr., S.: Perception of emotion: measuring decoding accuracy of adult prosodic cues varying in intensity. J. Nonverbal Behav. **22**, 89–107 (1998)

Brookes, T., Williams, D.: Perceptually-motivated audio morphing: Brightness. In: Audio Engineering Society Convention 122, Audio Engineering Society (2007)

Caetano, M., Rodet, X.: Independent manipulation of high-level spectral envelope shape features for sound morphing by means of evolutionary computation. In: Proceedings of the 13th International Conference on Digital Audio Effects (DAFx), vol. 21 (2010)

Cospito, G., de Tintis R.: Morph: Timbre hybridization tools based on frequency. In: Workshop on Digital Audio Effects (DAFx-98) (1998)
Coutinho, E., Cangelosi, A.: Musical emotions: predicting second-by-second subjective feelings of emotion from low-level psychoacoustic features and physiological measurements. Emotion. **11**, 921–937 (2011). https://doi.org/10.1037/a0024700
Daly, I., Williams, D., Hallowell, J., Hwang, F., Kirke, A., Malik, A., Weaver, J., Miranda, E., Nasuto, S.J.: Music-induced emotions can be predicted from a combination of brain activity and acoustic features. Brain Cogn. **101**, 1 (2015)
Deng, J.J., Leung, C.H.C.: Music retrieval in joint emotion space using audio features and emotional tags. In: Li, S., Saddik, A., Wang, M., Mei, T., Sebe, N., Yan, S., Hong, R., Gurrin, C. (eds.) Advances in Multimedia Modeling Lecture Notes in Computer Science, vol. 7732, pp. 524–534. Springer, Berlin (2013)
Frick, R.W.: Communicating emotion: the role of prosodic features. Psychol. Bull. **97**, 412 (1985)
Garrard, C., Williams, D.: Tools for fashioning voices: an interview with Trevor Wishart. Contemp. Music. Rev. **32**, 511–525 (2013)
Gobl, C.: The role of voice quality in communicating emotion, mood and attitude. Speech Comm. **40**, 189–212 (2003). https://doi.org/10.1016/S0167-6393(02)00082-1
Haken, L., Fitz, K., Christensen, P.: Beyond traditional sampling synthesis: real-time timbre morphing using additive synthesis. In: Analysis, Synthesis, and Perception of Musical Sounds, pp. 122–144. Springer, New York (2007)
Josephson, D.: A brief tutorial on proximity effect. In: Audio Engineering Society Convention 107. Audio Engineering Society (1999)
Juslin, P.N., Laukka, P.: Emotional expression in speech and music. Ann. N. Y. Acad. Sci. **1000**, 279–282 (2006). https://doi.org/10.1196/annals.1280.025
Kotlyar, G.M., Morozov, V.P.: Acoustical correlates of the emotional content of vocalized speech. Sov. Phys. Acoust. **22**, 208–211 (1976)
Le Groux, S., Verschure P.F.M.J.: Emotional responses to the perceptual dimensions of timbre: a pilot study using physically informed sound synthesis. In: Proceedings of the 7th International Symposium on Computer Music Modeling and Retrieval, CMMR (2010)
Mo, R., Bin, W., Horner, A.: The effects of reverberation on the emotional characteristics of musical instruments. J. Audio Eng. Soc. **63**, 966–979 (2016)
Olivero, A., Depalle P., Torrésani B., Kronland-Martinet R.: Sound morphing strategies based on alterations of time-frequency representations by Gabor multipliers. In: Audio Engineering Society Conference: 45th International Conference: Applications of Time-Frequency Processing in Audio. Audio Engineering Society (2012)
Pell, M.D.: Cerebral mechanisms for understanding emotional prosody in speech. Brain Lang. **96**, 221–234 (2006)
Sethares, W.A., Milne, A.J., Tiedje, S., Prechtl, A., Plamondon, J.: Spectral tools for dynamic tonality and audio morphing. Comput. Music. J. **33**, 71–84 (2009)
Williams, D.: Affective potential in vocal production. In: Audio Engineering Society Convention 139. Audio Engineering Society (2015a)
Williams, D.: Developing a timbrometer: perceptually-motivated audio signal metering. In: Audio Engineering Society Convention 139. Audio Engineering Society (2015b)
Williams, D., Kirke, A., Eaton, J., Miranda, E., Daly, I., Hallowell, J., Roesch, E., Hwang, F., Nasuto, S.J.: Dynamic game soundtrack generation in response to a continuously varying emotional trajectory. In: Audio Engineering Society Conference: 56th International Conference: Audio for Games. Audio Engineering Society (2015)
Wishart, T.: The composition of "Vox-5". Comput. Music. J. **12**, 21–27 (1988)

Chapter 4
Affectively-Driven Algorithmic Composition (AAC)

Duncan Williams

Introduction

This chapter introduces a working concept which a number of subsequent chapters will rely upon: Affectively-Driven Algorithmic Composition (or AAC). The reader should note that this is not related to perceptual data compression as in the Apple Lossless file format AAC. Instead it refers to a specific subset of interdisciplinary practices marrying sound design opportunities with emotional intent; a paradigm which is ideally suited to modern video game soundtracking practice. This chapter builds upon initial work reported in the ACM Computers in Entertainment journal (though in an online article, not a specific journal edition), in 2017 (Williams et al. 2017).

Call of Duty: Modern Warfare features 17 unique pieces of music, around 52:31 total duration, using 531 MB of storage. Players might spend upwards of 100 h playing the game, and are therefore likely to have heard each piece of music many times over. If a system like the AAC pilot evaluated here could be expanded upon to sustain or improve player immersion as well as improve emotional congruence, the benefits to the video game world would not simply be limited to a reduction in workload for composers, or to less repetitive sound-tracking for players—the amount of data storage required for soundtrack storage might also be significantly reduced (a stereo CD-quality PCM wave file takes approximately 10mb of storage per minute of recorded audio). Whilst data space is not as scarce a commodity in the modern world as it might have been in the days of game storage on cartridge or floppy disk, gaming is now increasingly moving onto mobile platforms (phones, tablets) with limited storage space, and online streaming is also a popular delivery platform for gaming. Therefore, a reduction in data by using AAC for sound-tracking

D. Williams (✉)
Digital Creativity Labs, University of York, York, UK
e-mail: duncan.williams@york.ac.uk

D. Williams, N. Lee (eds.), *Emotion in Video Game Soundtracking*, International Series on Computer Entertainment and Media Technology,
https://doi.org/10.1007/978-3-319-72272-6_4

could represent a significant and valuable contribution to video game delivery in the future.

This chapter outlines the design and implementation of one such system, and evaluates it in use in a real-world gaming study using a MMORPG (a massively multiplayer online role-playing game). In the case of the experiments documented here, the intention ultimately is to create a system for automated generation of new music which can both reflect and induce a change in the listener's affective state. As reported in Chaps. 2 and 3, the distinction between *perceived* and *induced* affective state is important: the affective state of the listener must actually change in response to the musical stimulus in order for the state to be considered induced.

Defining AAC

Algorithmic composition, and the large variety of techniques for computer automation of algorithmic composition processes, are well documented in the literature. Surveys of expressive computer performance systems such as that carried out by Kirke and Miranda (2009) also provide a thorough overview of the extensive work carried out in the area of emotionally targeted computer aided music performance. Rowe (1992) describes three distinct approaches in generic algorithmic composition systems: generative, sequenced, or transformative. Sequenced systems make use of pre-composed musical passages, which are subsequently ordered according to the algorithm. Generative systems create new musical passages according to particular rulesets (often, the selective filtering of random data). Transformative systems, the type evaluated in this chapter, take existing musical passages as their source material and derive new sequences according to various functions which might be applied to this source material (for example, a basic transformation might be to reverse the notes of a musical sequence—commonly referred to as a retrograde transformation).

Affectively-driven algorithmic composition (AAC) is a specific sub-category of this field which has come to light in the past decade. The idea is to use algorithmic techniques to expand upon musical ideas, specifically to generate new music with particular emotional qualities or affective intentions. More than reflecting emotions, a central application of generative music, and in particular AAC, is to develop technology for building innovative intelligent systems that can not only monitor emotional states in the listener, but also induce specific affective states through music, automatically and adaptively. This offers the potential to significantly enhance the players emotional interaction with a game. In order for such systems to be useful in the context of video game soundtracking, they must be capable of creating a theoretically infinite number of soundtracks without time constraints. One example of the utility of such techniques would be to avoid a situation where the player becomes exposed to the same exact piece of music multiple times, which can cause unexpected or unpredictable emotional responses (i.e., there is no saying when a particular piece of music may need to end or change in emotional content as this is dictated in real-time by the player). A disadvantage to looping-based approaches for gameplay soundtracking is the high amount of repetition involved. This can become distracting or worse, irritating, at transition points, which can have a knock-on negative effect on player immersion. The resolution of looping systems can be improved by adding divergent score 'branches' at narrative breakpoints within the soundtrack, which results in more complex, less repetitive musical sequences. However, the need to create the contributory score fragments in such a manner that they can be interchanged whilst maintaining the intended aesthetic-congruency with the narrative poses a significant challenge to the video-game composer. In simple terms, the over-arching challenge is that video game music can get repetitive and thereby ruin player immersion, but composing large quantities of music with particular moods and emotions is not practical for most games, both in terms of storage on the media (whether that be disc, cartridge, or simply bandwidth in, for example, online streaming games), and also in terms of human cost (i.e., that of the composers time when constructing large numbers of interchangeable musical sequences). Thus, the adaptability of a branching system to emotional responses for these purposes is somewhat compromised. Repetition in video games soundtracking has previously been shown to be detrimental to immersion (Brown and Cairns 2004). This is a central challenge to AAC, because repetition is part of what gives us structure, expectation, and understanding when listening to soundtracks.

Prototype Design and Implementation

A prototype AAC system which incorporates a range of musical features with known affective correlates; tempo, mode, pitch range, timbre, and amplitude envelope, was developed and subjected to a perceptual evaluation which will be documented and explained here. A series of affective mappings (musical features with emotional responses) was drawn from the literature—see Williams et al. (2014) for the full literature review—and implemented in an artificial intelligence driven AAC system. Initially a single musical feature, rhythmic density, was evaluated (Williams et al. 2015). This feature can contribute to perceived tempo (which, as mentioned above, has been suggested to be well correlated with affective arousal) as well as other sub-features (for example articulations like staccato or legato performance). This system was subsequently expanded to include the larger matrix of musical features listed above. It works by exploiting specific variations in each of these musical features to imply different affective states in newly generated music samples, according to a 3 × 3 Cartesian grid across a 2-D affective space based on the circumplex model (Russell 1980). These mappings utilize a time series of varying proportions of these musical features, intended to evoke particular affective states on the 2-Dimensional affective space.

Seed material is generated by means of a 16-channel feed-forward artificial neural network (ANN). ANNs have previously been used in algorithmic music systems in Bresin (1998). The ANN in this case is trained on 12 bars of piano music in C

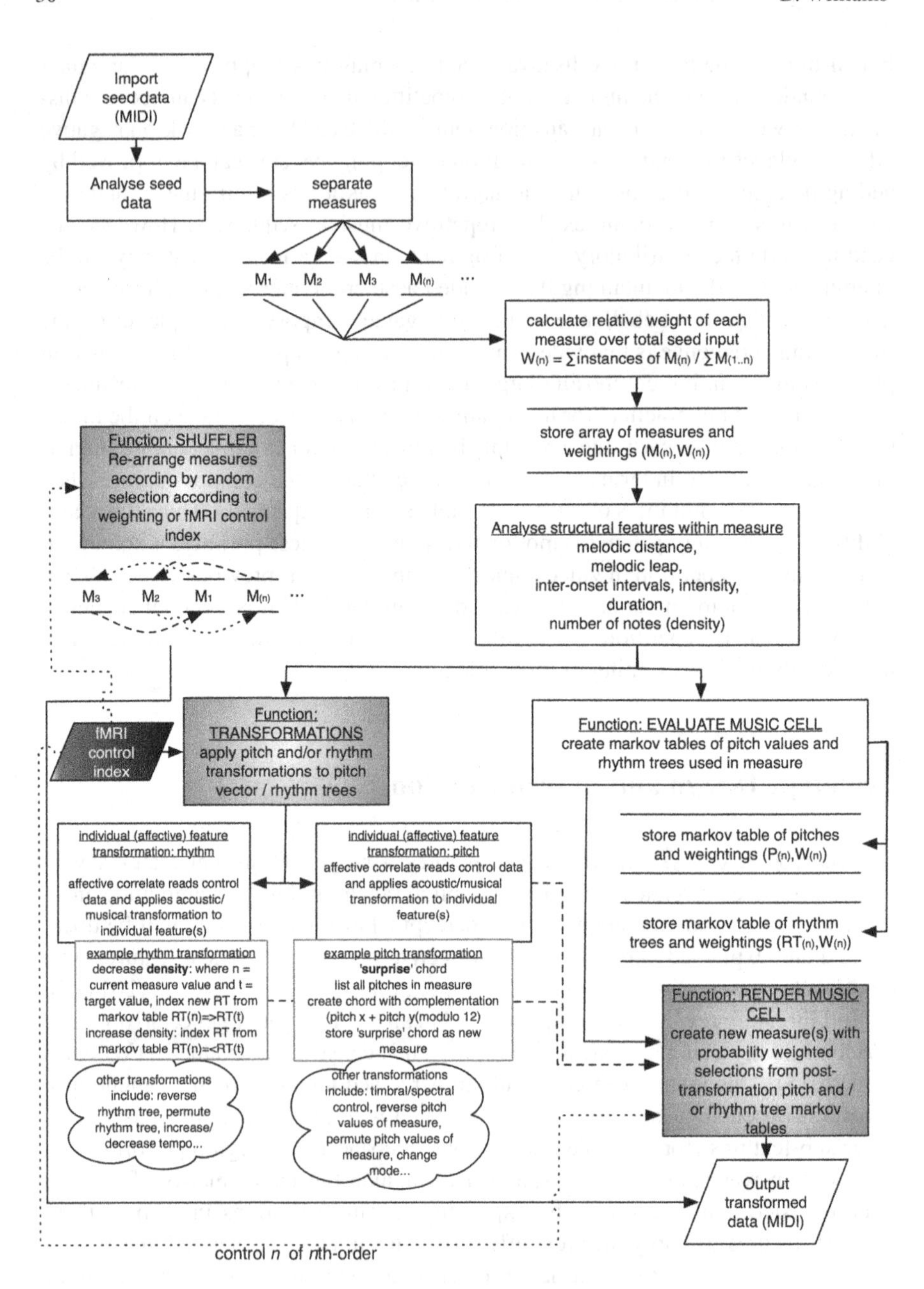

Fig. 4.1 Schematic of a generation system employing emotionally-driven transformations in the creation of new (MIDI representations) of musical structures

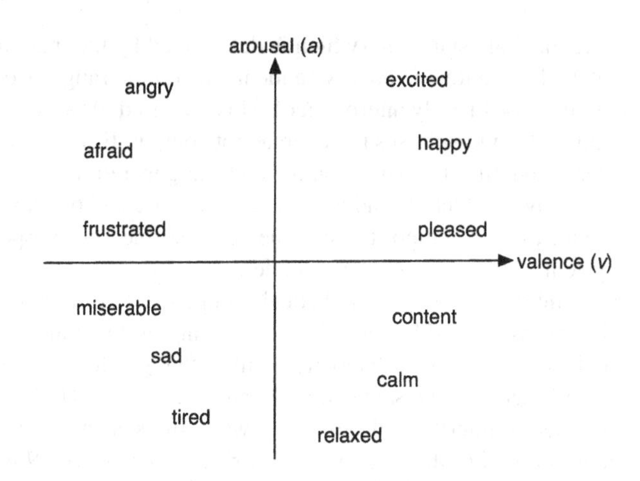

Fig. 4.2 2-Dimensional affective space used for targeted stimulus generation, showing Cartesian co-ordinates in arousal (emotional intensity) and valence (positivity of emotion), after the circumplex model of affect (Russell 1980)

major at 120 bpm, as shown in Fig. 4.1. Music is input as MIDI data and output in the same format. The ANN hidden layers are trained according to the weighting of each musical feature (*tempo, mode, timbre, melodic range, dynamic range*) from the seed material (off line). There is one layer for each half step in an octave (16 notes, rather than the 12 on a piano, i.e., we allow for both Bflat and Csharp etc), where p is the input vector, a is the output vector, and w is the weighting for each input as determined by the transformation/activation function (essentially, transformation f creates weighting w which then acts on input p to create output a). For each epoch:

1. Initialize the neural network, set weights w_{ij} for each $_j$th node for each area of the 2-dimensional emotion space
2. Input neurons x_{ij} from target set of emotional correlates (musical features from matrix)
3. Apply transformation/activation function f, f(x)
4. Change weights w_{ij} of nodes in hidden layers
5. Output layer (generate score for performance)

This approach has been used for many supervised learning applications, for example Carpenter et al. (1992).

The seed pool of material is then further transformed according to the affective mapping of five musical sub-features: tempo, mode, pitch range, timbre, and amplitude envelope. Specific variations in each of these musical features are used to imply different affective states in the generated material according to a 3 × 3 Cartesian grid across the 2-D space, as shown in Fig. 4.2, ranging from low valence and low arousal in the bottom left of the space, to high valence and high arousal in the top right of the space. Tonality is manipulated by selecting from a variety of scales and pitch quantizing the pitch class accordingly (harmonic minor, chromatic, major, pentatonic) and the generator than transposes that scale type so that it can cover all keys. Thus there is no additional requirement to list modes separately because as far as pitch class content is concerned they are identical to the parent scale. In other words, the only difference between C Maj and G mixolydian is which note the scale starts on, (which is determined by the incoming MIDI notes), as the pitch class content is otherwise identical. Tempo ranges from 90,120,150 bpm (but these can be linearly interpolated). This is carried out as a number of initial messages sent to the pitch classes to generate note output. Envelope is a simple MIDI parameter amounting to sustain on/off and the generation of the legato/staccato effect, respectively. Pitch spread is a range between 2 and 6 octaves across the piano keyboard, as with tempo, these values can be linearly interpolated on the fly by the system when it is in generate mode.

Timbre variation is simplistically implemented, we assume that brighter, harder timbres as created by more intense playing on the piano are mapped to the top 42 MIDI cc messages for 'velocity', with softer, gentler timbres in the bottom 42 MIDI cc messages for the same control value. As with pitch and tempo these values can be linearly interpolated on-the-fly when the system is in generate mode, and the ranges for all of these can be over-written when the ANN is offline.

The generated musical samples are performed by a synthesized piano voice. Co-ordinates with higher arousal generally include a larger pitch spread (range of notes), faster tempo, and harder timbres, whilst co-ordinates with higher valence generally utilize a major key. In this system, a Cartesian coordinate of (valence [1:3], arousal [1:3]) is used to refer to a given combination of the five musical features, (*tempo, mode, timbre, melodic range, dynamic range*). Thus a co-ordinate of (1, 1) would refer to low valence and low arousal, the lowest corner to the left of the space. This co-ordinate would force the transformation algorithm to create stimuli incorporating a slow tempo, a minor key, a soft timbre (on a piano, the timbre of the performance can be manipulated using dynamics markings, where perceptually harder sounds are achieved with louder performance dynamics), an amplitude envelope with considerable legato, and a spread of pitch values, which are comparatively lower than those of the rest of the generated pool. An example is given in Fig. 4.3.

Prototype Evaluation

A tri-stage listener evaluation was used to inform two levels of subsequent adjustment to the feature mappings wherein the size and spread of affective responses was gradually increased by deliberate manipulation of the combinations of musical features in the affective mappings. Listener evaluation of a prototype AAC system confirmed that affective targeting via novel musical material is possible if the material is generated by means of an underlying matrix of musical features with documented emotional correlations. A good spread of emotional responses was achieved, broadly comparable to the spread in a real-world stimulus set, but the real-world stimulus set was not directly evaluated against the generated stimulus set, and thus the generator should be considered as conceptually distinct from the real-world stimulus set, a direct comparison would require further experiment. The basic technique for generating music described here is not novel in and of itself. Many other successful implementations of different types of probabilistic analysis and

Fig. 4.3 Eight bars of a generated musical figure with a target affective co-ordinate of (v1, a1), low valence and low arousal. Note that the musical features include a lower pitch spread (generated material is in the bass clef) but retain some of the features of the training material (such as the single instance of a triplet, bar 1) and instances of ¼ note rests as well as a mixture of mainly 1/8th and 1/16th note rhythms. The key signature implies C minor

generation exist, including Markov models (Casey 2001; Visell 2004) and recurrent neural networks (Bown and Lexer 2006), amongst others. However, the combination of generative AI techniques with affective feature mapping does offer a novel way of creating and interacting with emotionally expressive music, with enormous potential in the context of creating opportunities for musical communication, for example with a virtual co-performer in games like *Guitar Hero* (Williamon and Davidson 2002). The field of AAC is in its infancy and therefore there are no major precedents for evaluation of such a system. Methods for evaluating the utility of music creation systems, particularly algorithmic composition are the subject of much debate, including in some cases debate as to the question of authorship of music created by such systems (Dahlstedt 2001; Eigenfeldt 2011). This of course depends on the type of game being soundtracked, and the purpose of the soundtracking. Nevertheless, combining these techniques with deliberate affective targeting, in response to a real-time game performance, does compound these issues. Therefore, beyond perceptual calibration of the musical feature emotion space mapping, and other issues for practical implementation in the generative system, there remained a significant amount of further work to address in devising practical and robust evaluation methodologies for AAC systems in the context of real-time game playing.

Testing in Gameplay

A quest (section of gameplay) from *World of Warcraft* (an MMORPG), was marked up with various affective targets as meta-tags (for example, fighting scenes were tagged with {v1, a3}, or angry). Stimuli corresponding to the affective meta-tag were created using the AAC prototype and streamed during gameplay via a dedicated audio playback engine built using Max/MSP. Participants were then asked to complete the quest three times; once with the original musical soundtrack, once with a soundtrack provided by the algorithmic composition system, and once with no musical accompaniment at all (sound effects were still used, for example, action sounds). Each playback was recorded so that the stimulus selections could be repeated for subsequent affective evaluation. In a pre-evaluation questionnaire, the majority of the participants reported that they enjoyed listening to music, while only 45% of them had experience of composing or performing music. 72% of participants reported that they enjoyed playing or watching video games, although 45% of participants answered that they only spend 0.5 h a week watching or playing video games. Participants were asked to rate emotional congruence and immersion after each playthrough, using a 11-point Likert scale presented via an interactive web-browser form. Short definitions for both terms were included in the pre-experiment instructions for participants.

Having evaluated emotional congruence and immersion for each of the musical playthroughs, participants were also asked to rate the perceived emotion of each stimulus that they had been exposed to in the generative soundtrack playthrough, using a two-dimensional space labeled with the self-assessment manikin (Bradley and Lang 1994) showing valence on the horizontal scale and arousal on the vertical scale, allowing both valence and arousal to be estimated in a single rating. Valence was defined to the participants as a measure of positivity, and arousal as a measure of activation strength in a pre-experiment familiarization and training stage. This participant interface was also implemented in Max/MSP. Participants undertook the experiment using circumaural headphones in a quiet room with a dry acoustic. The exact duration of playthroughs varied on a player-by-player basis, from 4 min to a ‘cut off’ of 10 min. A variety of affective states may be entered in each playthrough, depending on the actions of the player, for example exploration, fighting, fighting and being victorious, fighting and failure, evading danger, and interacting positively with other (non-player) character avatars.

Some participants anecdotally reported that the generated music seemed more responsive to the gameplay narrative, and therefore added an extra dimension to the gameplay (in other words, they felt "in control" of the music generation system). This suggests that this type of system might have possible applications as a composition engine for pedagogic, or even therapeutic applications beyond the world of entertainment. In general terms, players found the generated soundtrack more emotionally congruent with the gameplay than that of the original soundtrack, and that this marked improvement was both consistent across all participants, and statistically significant. This was a promising result for the AAC system. However, the mean immersion decreased in the ratings of the generated soundtrack playthrough in comparison to the original soundtrack playthrough. This suggests that player immersion was consistently reduced in the generated soundtrack playthrough.

The average ratings for emotional congruence and immersion in playthroughs with the original soundtrack and the generated soundtrack suggested that although the results were close, the difference between original and generated ratings in emotional congruence and immersion is significant. This suggests that the improvement between emotional congruence in the original soundtrack and the generated soundtrack is consistent. Participants were not asked to rate the emotional congruence of the soundtrack in the silent playthrough. The infancy of the field at the time of conducting these experiments means that there are few precedents for appropriate evaluation paradigms, and to test for normality would take hundreds of trials, which might also have a knock-on effect on the participants involved (repeatedly undertaking the same section of gameplay might have an impact on immersion for example).

The generated soundtrack was performed via a single synthesized piano timbre, which is in a sharp contrast to that of the original soundtrack, which consisted of a fully orchestrated symphonic piece. Nevertheless, participants seemed to consistently find that the generated music matched the emotional trajectory of the video gameplay more congruently than the original musical soundtrack. The majority of participants also reported that the playthroughs which were accompanied by a musical soundtrack were more immersive than the silent playthrough (which featured sound effects but no musical accompaniment). However, there was also a marked decrease in reported immersion with the gameplay in the generated soundtrack playthroughs, perhaps because of the lack of familiarity and repetition in the generated soundtrack. This could be usefully addressed in further work by evaluating repetition of themes and with generated music across multiple different games with the same player in future. Another possible explanation for this increase in emotional congruence but decrease in immersion might well be the orchestration of the AAC system, (solo piano). The original soundtrack might be lacking in emotional variation, but it offers a fullness and depth of instrumentation which is not easily matched by a single instrumental voice. This is a challenge which could be evaluated in future work by creating a piano reduction of the original gameplay score and subjecting it to subsequent evaluation, or by developing the generative system further such that it can create more fully realized pieces of music using multiple instrumental timbres (this alone presents a significant challenge to algorithmic composition). Together, this suggests a strong argument that adapting this type of system to individual responses would be a useful avenue for further work, for example by calibrating the musical feature-set from the generative algorithm to each individual player before commencing, or by using bio-feedback data (e.g., from a brain cap, heart-rate monitor or other biosensor, as outlined in Chap. 9) to calibrate the musical feature-set on a case-by-case basis to attempt to reduce inter-participant variability. This type of testing would have implications on the number of trial iterations required, and the trial time that participants were required for (at present, the maximum test time was approximately 30 min, after which point listener fatigue might be considered a factor). Our hypothesis would be that emotional congruence, and likely immersion, would go down in such a case, but this has not yet been tested and remains an avenue we intend to explore in further work.

Conclusions

Using an AAC system to create music according to emotional meta-tagging as part of a video game narrative has clear practical benefits in terms of composer time and hardware/software requirements (file storage and data transmission rates), which could free up processing and storage space for other game elements (e.g., visual processing). This type of system might also therefore be beneficial to mobile gaming platforms (smart phones etc.), where space is at more of a premium than on desktop or home gaming environments. In this work the particular combination of these features has been explored and adjusted in response to listener evaluation to successfully exploit a large portion of a 2-Dimensional emotion space, but the complex nature of the inter-relationship between these musical features, and the subsequent affective responses that might be the subject thereof, remains an area of considerable further work.

Within the constraints of the test paradigm, this pilot study suggests that emotional congruence could be improved when participants played with a soundtrack generated by the affectively-driven algorithmic composition system. However, player immersion was consistently and significantly reduced at the same time. It might be possible to seek explanation of this in the instrumentation and orchestration of the generated music, but further work would be required to establish the reason for this reported reduction in player immersion, before tackling the problem of developing an AAC system with fuller orchestration, which is in and of itself non-trivial. These algorithmic composition techniques are still in their infancy and the likelihood of replacing a human composer in the successful creation of complex, affectively-charged musical arrangements is minimal. In fact, as the system presented here (and others like it) require training with musical input, this evaluation suggests that in the future composers working explicitly with video game soundtracking might use this type of system to generate large pools of material from specific themes, thereby freeing up time to spend on the creative part of the composition process.

Participant agreement with the affective meta-tagging used to select musical features as part of the generative system was good, though significant inter-participant variability suggested that either the musical feature set needs further calibration (which would require specific affective experiments), or that a generalized set of affective correlates as musical feature sets is not yet possible. Another solution might be to calibrate this type of generative music system to the individual, using a mapping of musical features documented here in order to attempt to target specific emotional responses in the generated soundtrack. In the future, this could be driven by bio-sensors such as the electroencephalogram (as in the emerging field of brain-computer music interfacing), or by more traditional biosensors such as heart rate sensors or galvanic skin response. Should such a system be adaptable to real-time control by means of biophysiological estimates of affective state, a feedback driven AAC system could be created for continuous monitoring and induction of target affective states in the player (e.g., for therapeutic means) in the future.

Broader musical structure, including thematic variation and repetition, is not addressed in this pilot system beyond the matching of a single emotional trajectory according to the player character. Thus, this evaluation should not be generalized beyond MMORPG games, and in future would be best expanded to an evaluation across a wide range of games for each participant. Moreover, the possible positive influence of repetition on player immersion should not be discounted. Structural composition techniques remain challenging for all automatic composition systems (Edwards 2011) and as such present a fertile area for continued investigation beyond the scope of the work presented here.

References

Bown, O., Lexer, S.: Continuous-time recurrent neural networks for generative and interactive musical performance. In: Applications of Evolutionary Computing, pp. 652–663. Springer, Cham (2006)

Bradley, M.M., Lang, P.J.: Measuring emotion: the self-assessment manikin and the semantic differential. J. Behav. Ther. Exp. Psychiatry. **25**, 49–59 (1994)

Bresin, R.: Artificial neural networks based models for automatic performance of musical scores. J. New Music Res. **27**, 239–270 (1998)

Brown, E., Cairns P.: A grounded investigation of game immersion. In: CHI'04 Extended Abstracts on Human Factors in Computing Systems, pp. 1297–1300. ACM, New York (2004)

Carpenter, G.A., Grossberg, S., et al.: A self-organizing neural network for supervised learning, recognition, and prediction. IEEE Commun. Mag. **30**, 38–49 (1992)

Casey, M.: General sound classification and similarity in MPEG-7. Organ. Sound. **6**, 153 (2001). https://doi.org/10.1017/S1355771801002126

Dahlstedt, P.: A MutaSynth in parameter space: interactive composition through evolution. Organ. Sound. **6**, 121 (2001). https://doi.org/10.1017/S1355771801002084

Edwards, M.: Algorithmic composition: computational thinking in music. Commun. ACM. **54**, 58–67 (2011)

Eigenfeldt, A.: Real-time composition as performance ecosystem. Organ. Sound. **16**, 145–153 (2011). https://doi.org/10.1017/S1355771811000094

Kirke, A., Miranda, E.R.: A survey of computer systems for expressive music performance. ACM Comput. Surv. **42**, 1–41 (2009). https://doi.org/10.1145/1592451.1592454

Rowe, R.: Interactive Music Systems: Machine Listening and Composing. MIT Press, Cambridge (1992)

Russell, J.A.: A circumplex model of affect. J. Pers. Soc. Psychol. **39**, 1161 (1980)

Visell, Y.: Spontaneous organisation, pattern models, and music. Organ. Sound. **9**, 151 (2004). https://doi.org/10.1017/S1355771804000238

Williamon, A., Davidson, J.W.: Exploring co-performer communication. Music. Sci. **6**, 53–72 (2002)

Williams, D., Kirke, A., Miranda, E., Daly, I., Hallowell, J., Weaver, J., Malik, A., Roesch, E., Hwang, F., Nasuto, S.: Investigating perceived emotional correlates of rhythmic density in algorithmic music composition. ACM Trans. Appl. Percept. **12**, 8 (2015)

Williams, D., Kirke, A., Miranda, E., Daly, I., Hwang, F., Weaver, J., Nasuto, S.: Affective calibration of musical feature sets in an emotionally intelligent music composition system. ACM Trans. Appl. Percept. **14**, 1–13 (2017). https://doi.org/10.1145/3059005

Williams, D., Kirke, A., Miranda, E.R., Roesch, E., Daly, I., Nasuto, S.: Investigating affect in algorithmic composition systems. Psychol. Music. **43**, 831 (2014). https://doi.org/10.1177/0305735614543282

Chapter 5
An Auto-Ethnographic Approach to Creating the Emotional Content of Horror Game Soundtracking

David Bessell

Introduction

The study of creative process in any field of art is something that presents a kind of resistance to more usual academic and scientific modes of examination. Or to be more precise, certain aspects of the creative process present difficulties for these modes of investigation.

> Is it appropriate to investigate creative processes from a mechanistic perspective or do they involve subjective elements which cannot – in principle - be investigated from such a perspective? (Gonzalez and Haselager 2003)

In particular the study of music and sound compounds this problem as aspects of the processes by which music exerts such a strong emotional and psychological pull on an audience are not currently fully understood. Add to this the complexities introduced by the use of music in the non-linear landscape of the computer game world and the difficulty of examining how video game soundtracks are constructed to engage players at an emotional level multiplies exponentially. In this context it is useful to consider film soundtracks. Bordwell has suggested

> … we cannot stop the film and freeze an instant of sound, as we can study a frame to examine mis-en-scene and cinematography. Nor can we lay out the soundtrack for our inspection as easily as we can examine the editing of a string of shots. In film, the sounds and patterns they form are elusive. This elusiveness accounts for part of the power of this technique:… (Bordwell 1979)

Actually this is not strictly technically true, a written score can show us the micro detail and larger scale structure of film music and in relatively recent times computer based recordings can also display a variety of visual and audio information that

D. Bessell (✉)
ICCMR, Plymouth University, Plymouth, UK
e-mail: david.bessell@plymouth.ac.uk

D. Williams, N. Lee (eds.), *Emotion in Video Game Soundtracking*, International Series on Computer Entertainment and Media Technology,
https://doi.org/10.1007/978-3-319-72272-6_5

allows a moment by moment reading of the audio content of a film. However Bordwell's comments start to make more sense when applied to the non-linear context of a game soundtrack. In the non-linear context we are cast adrift on a sea of more or less indeterminate audio possibilities which are not definitively fixed in time.

If we allow that the above is the case then an auto-ethnographic approach from the perspective of the game soundtrack composer is well suited to provide some further information on what may be occurring in the emotional engagement engendered by games music. At the very least it may more precisely delineate the areas of creative sound design and composition for games that are currently poorly understood and may benefit from further investigation. To that end this article forms a case study of the process of game soundtrack creation based on the authors ongoing experience of soundtrack creation for the game "Deal With The Devil" which is currently in production (Deal with the Devil, Round Table Games unpublished). This builds on some limited precedents for auto-ethnographic studies of the process of game sound creation (Baystead 2016) but these are mainly concerned with other aspects of the process such as the pragmatic hardware limitations placed on the soundtrack composer.

Breaking the Problem Down

I think it would be fair to say that musicians themselves have not historically been terribly concerned with a self reflexive examination of the creative process at the point at which it occurs. Indeed it is not uncommon to encounter a specific reluctance to undertake any such examination based on the idea that this creative process is somehow mysterious and elusive, prone to evaporation under too harsh a glare of logical analysis. So where to begin?

Self evidently the emotional effects of music in any context are not just down to one or two things but are more likely due to the interaction of a variety of competing and complimentary factors. However we can still usefully broadly divide the mechanisms that drive emotional processes within the music into some discrete categories. These categories that outline some of the important means by which music can interact emotionally with an audience include: Associative aspects, learned skills and intrinsic properties of sound. This last category 'intrinsic' or inherent structural properties of sound is somewhat controversial and there are live debates within the world of composers on whether this property exists or is a subset of the learned category. For my own part as a composer I experience certain aspects of sound such as the harmonic series as intrinsic and therefore I will deal with them as such here. General examples of these categories at work in music might include: Associative, music that evokes by its context a strong sense of historical time or geographical place such as the popular music of the 1930s or Indian classical music. Learned, the emotional associations commonly attributed to various musical constructs such as the sadness of the minor scale, or the positivity of the major scale. Intrinsic, might for example include the sense of none specific emotional arousal associated with fast tempi.

> Through intertextual references to audio-visual idioms from other media, game soundtracks deploy player literacy for their immersive effect: it is because gamers recognise certain composing styles that they are able to interpret gaming events and feel involved in game-play… (Van Elferen 2016)

The role of these three categories in relation to audience emotion is further compounded when music appears in a relation to picture because the pictures also have associative, learned and intrinsic qualities of their own. Furthermore the introduction of non-linear elements in the relation between music and pictures also complicates the possibilities, as there may be a variety of more or less unforeseen juxtapositions of music and picture. This is of course the situation in the majority of video game soundtracks where the order of events can be at least partially determined by the player at the local level. The juxtaposition of particular music and particular pictures can feed off each other in different ways, they can support or reinforce each other, they can contradict each other and they can refer to different events that have either happened or will happen (temporal displacement). They can also maintain different independent narratives at the same time. The soundtrack can be diegetic or non-diegetic.

> While they (the player/audience) are still, in a sense, the receiver of the end sound signal, they are also partly the transmitter of that signal, playing an active role in the triggering and timing of these audio events. Existing studies and theories of audience reception and musical meaning have focused primarily on linear texts. (Collins 2008)

> …the audience becomes an (inter)active component of meaning making in the consumption of any media text. (Collins 2013)

The game *Rez* (Sega 2001) is a particularly interesting example of the blurred line between an interactive score and the game players having a role in creating the soundtrack in real time.

The concept of immersion is also key in generating effective emotional responses from the game player, if the player is not fully engaged with the game play then any musical content designed to elicit emotion may well go unheeded. It is the job of the soundtrack composer to successfully navigate this complex creative space defined by the interactive computer games context.

These varied and shifting combinations of sound and picture provide rich and complex possibilities for the creation of an emotionally engaging and immersive experience for the game player. The skill of the game soundtrack composer lies in being able to synthesise these complex variables into an effective and immersive emotional experience that enhances the game play. We have only to turn down the sound while playing a game to demonstrate how heavily dependent the emotional aspect of the experience is on an effective soundtrack. In Jorgenson's study where the sound was suddenly switched off while playing World of Warcraft

> Petter describes it as being left completely in the dark, while Nils compares it to losing a leg. Similarly Anders feels as if he has become blind. (Jorgenson 2009)

On Creativity

At the base of composing an effective soundtrack that emotionally engages a game player lies the creative process itself. It is my experience that musical creativity can appear quite different from the perspective of the creator than it does from the point of view of an external observer. Subjectively in my own working processes on the development work for *Deal with the Devil* there are several broad aspects at work. Firstly there are more overt applications of musical craft skills based on the western musical tradition of scales, modes, chords, counterpoint, motif, tonality, atonality and so on. Secondly come the less conscious but no less important sorting of the possibilities created by this artisan knowledge based on hard to quantify notions such as appropriateness for the situation, best fit, emotional resonance, most elegant solution, most effective solution, etc. There are also contributions from more hidden psychological processes such as random juxtaposition of dissociated ideas, intuitive connections between things not previously or normally associated that can give rise to new emotionally effective soundtrack solutions and novel responses. These less overt processes can also contribute to redefining the constraints of the creative space in which the work is realised. Without these more hidden psychological processes soundtrack work and music in general would be reduced to just regurgitating recombinations of previously known examples and nothing substantially new would ever emerge. An element of novelty or surprise is an important factor in generating particular types of emotional response, perhaps particularly so in the horror genre.

The intuitive element is particularly interesting as it implies a set of hidden (occulted) judgments as to what is appropriate and effective selected from the infinite number of random elements that could be combined. For it to be a useful creative strategy it must rely on something which is more or less in common between the composer and the player in order for emotional connections to be made. This implies an element of two way communication at work but also in the most creative situations an element of revealing new possibilities based on a shared commonality but adding to it. What exactly is being communicated is a philosophical question beyond the scope of this chapter but it could be loosely described as a process of recognition in some form between the audience and the composer. In other words the process of effective emotional communication in a soundtrack is at least partly dependent on the composer having an emotional language and experience in common with the audience as well as being able to communicate that effectively.

During the process of soundtrack creation the creative processes outlined above interact with each other and are in a constant state of flux. At different times one or the other may be uppermost. So for example, when I experience a sense of creative flow the rule based constraints manifested as learned, craft based, musical knowledge are so deeply assimilated as to mostly disappear from my conscious thought and are replaced by an illusion of direct unmediated creative manipulation of sound based on my artistic goals at that particular moment. At other times I may think more overtly about what musical scale, mode or rhythm I need to enhance a specific atmosphere or mood.

One final aspect of the creative process that seems to me to be a precondition for the creation of effective emotional immersion for the game player is the idea of constraints. The game context along with the specific brief and the learned musical tools all contribute to forming the boundaries of a particular creative space within which the soundtrack is realised. The interactive game space is particularly interesting in its implications for the composition of music because its non-linear properties remove one of the fixed constraints that have traditionally bounded the area of musical creativity in the western style. Notwithstanding some forays into non-linear form by the modernist classical avant garde, music has traditionally been an art realised on a fixed linear time scale and the removal of this constraint is likely to open up new avenues for creative innovation. The classical world presents models for this renewal of constraints as an important part of innovation in music.

> Schoenberg's creativity consisted not just in the rejection of a constraint but also in the generation of new ones which allowed new forms of musical composition. (Gonzalez and Haselager 2003)

It has almost become a cliché of discussions on creativity that constraints of some sort are a necessary pre requisite for getting started on a creative endeavour. The infinite possibilities need to be narrowed down to prevent the creative paralysis often associated with trying to parse an unmanageably large set of options at each specific creative juncture.

Examples of These Concepts at Work in *Deal with the Devil*

The brief for this particular game evolved (and is still evolving) as work on the game progresses. Perhaps unusually for game development, music and sound design were called for at a very early stage in the process. The first meeting with the game developers took place at the point at which only mood boards, a rather sketchy level storyboard, and some very limited concept art existed. The role of the music at this stage was to give the visual artists some sense of atmosphere and emotional cues to work from. This in itself is interesting as in this case the music began to colour and inspire the emotional content of the visual component of the game right from the outset. It suggests even visually biased artists look to sound in general and music in particular for emotional cues. Without any overt comment being made it seemed that music and sound design were by default chosen by the developers as the most relevant emotional tools with which to bootstrap the creative process.

The actual initial brief for the overall game was as follows—the game was to be in the horror genre, based very loosely on the fantasy literary work of H. P. Lovecraft. It was to be set in the 1930s and will eventually include different episodes which might take place in a variety of geographical locations. The main active agent was to be female who would undergo a transition of character or journey into madness depending on the in game choices made by the player. Game play would involve a number of well tried elements such as puzzle solving, adversarial situations and exploration of environments. There would be particular objects that would need to be found to progress the game play. In short many of the staples of the horror game genre. Some additional more specific details were communicated before audio work commenced for the first game level. Discussions were had about using a period piece of popular music from the 1920s or 1930s as the game theme and titles sequence. The track "Keep Young and Beautiful" by Abe Lyman 1933 was chosen by the game developers. During these initial discussions the horror element was further emphasised culminating with the verbal brief for one chase sequence being communicated as "scare the **** out of them"! Following this initial conversation a slightly more detailed but still very sketchy storyboard was supplied with approximate timings of the musical segments along with a list of sound effects and sound design elements required. Apart from this the musical brief was left very open with little in the way of specific detail, which from the creative point of view left me free to explore the appropriate sound world.

This initial brief went quite a way towards bounding a workable creative space in which to commence compositional work. As soon as I began to formulate more concrete ideas though I added further constraints to narrow down the options still more. These self imposed constraints included decisions about the musical styles that might be employed to enhance game immersion, atmosphere and audience identification. For example a mix of styles encompassing ragtime, the classical avant garde and some electronic elements were chosen. Even these initial decisions have relatively complex motivations and consequences for how emotional responses were built in to the unfolding soundtrack. To mention a few, the ragtime style has associative qualities that place it in the relevant period. Played on an upright 'thumbtack' piano it also brought associations of a somewhat decadent shadowy nightclub milieu associated with that historical period. The classical avant garde references made use of the intrinsic qualities of high levels of musical dissonance to create musical tension and high levels of musical change to signify chaotic out of control situations. The electronic elements were used for the contemporary associative qualities that they would bring for a typical horror game audience. The more intrinsic atmospheric and otherworldly qualities of electronic sound design were also judged as potentially useful. Whilst these decisions were initially conceived as constraints they also quickly became a nexus for new creative juxtapositions of musical elements that would not normally be heard in the same context.

Game Level 1

To throw further light on the decisions and techniques used to create the none too subtle emotional impact required for the horror game genre there follows a more in depth account of the creation of the soundtrack for a particular game section. This first game section involved the transition from the pre recorded linear game titles, to live interactive game play in a graveyard. The graveyard was to contain areas that needed to be explored and contained another character in addition to the main protagonist. Interaction with the environment and dialogue with the other character were required of the player in order to acquire knowledge and tools with which to progress to a mausoleum. Once in the mausoleum the game play involved unlocking a sarcophagus and unleashing a Lovecraftian horror. Action follows in which the main character, Amelia, tries to escape chased by this nameless winged horror. The soundtrack had to be able to represent what happens if Amelia was caught by the chasing horror and the level would end with a transition to another autonomous area of the game map once Amelia had been caught by the horror and had died, escaped or perhaps been possessed? The brief for this section also specified that the graveyard was geographically located in the UK. Interestingly the game developers wanted the Lovecraftian horror to be represented at least initially, primarily with sound. This was a consideration based on pragmatic decisions about developing in game 3D visual resources but also informed by the lineage of effective horror movie practice in which the 'monster' is more psychologically effective if less is actually shown on screen.

The first creative challenge was how to generate the appropriate emotional atmosphere of decadence and creepy foreboding from the 1930s title theme song which was in a major key and quite uptempo with a jaunty mood. Interestingly research into musical styles indicated that a great deal of the popular music of the 1920s and 1930s was uptempo, jaunty and clearly intended for dancing. Historically the social milieu of this period has been associated with a kind of hedonistic decadence so to an extent this came with the associative qualities of the given theme song. I decided that it would be interesting and more effective to sharply contrast the up tempo mood of the opening song with a transition to musical material derived from the song melody but transposed into a minor key and at a much slower tempo. This transition would colour the emotional response in a suitable way once the game play got under way. Unremitting musical tension and foreboding might in this context become self defeating after a certain time span and I decided that more intensely dark emotional responses could be had from the player if they were dropped from a bright sunny place (titles) into a deep dark abyss (game play). The transition from bright and jaunty to dark and foreboding was accomplished by the visual conceit of showing an old wind up phonograph slowing down at the end of the titles sequence. For this particular image the decision to progressively slow down the audio in the soundtrack creation had driven the visual inclusion of the wind up phonograph. This is an illustration that the process of creating effective emotional immersion was a somewhat circular one that involved back and forth dialogue with the visual artists and coders at work on the game play. As the old phonograph winds down the titles music slows to a funereal pace. As a result the pitch drops to a growl over which is introduced a fragmented minor version of the melody played on vibraphone. The vibraphone is one of the instruments chosen for its associations with the period, having been invented in the late 1920s. Conveniently it also blends well with more contemporary electronic instruments such as analogue synthesizers which I wanted to use to create other worldly textures and atmospheric sound design as a backdrop to the less frenetic exploratory game play that would follow the titles.

The appearance of the minor vibraphone melodic motif coincides with the first appearance of the main protagonist Amelia with which the game player is intended to identify. The player plays the game through the agency of her character. This seemed an ideal way to introduce a musical motif that could be associated in later game play with the main character. The Wagnerian concept of leitmotif has perhaps become a little too well worn in the world of film soundtracks so I decided to use it sparingly in this game but nevertheless I could not resist it from time to time as it is so effective at tying specific musical events to a character. Without any overt discussion on this topic the game developers picked up on this leitmotif aspect and began to incorporate it in the publicity trailers for the game. In this context the motif began to assume the role of theme for the game in general (*Deal with the Devil* 2017).

The first section of interactive game play consists of Amelia/the player exploring the environment of the graveyard at night. This called for a musical bed or underscore of indeterminate length that would heighten the horror mood of nocturnal unease and foreboding. For me one thing that is guaranteed to break player immersion in a game is an inappropriate transition from one section of music to another often caused by an exact repetition of a previously heard musical loop. To avoid this I created three musical layers of differing lengths which could be looped independently, which when combined would create an evolving atmospheric texture that could be continued for any needed length without the repetition becoming too obvious. Layer one was slow moving atmospheric and electronic with occasional brief references back to the slowed down growling version of the opening title music. This layer was relatively harmonically relaxed, loosely minor but drifting with no fixed sense of a tonic note. More other worldly than tense.

The second layer was some rather quiet high pitched dissonant sustained orchestral strings that were intended to create subdued tension with close chromatic musical intervals (Major and minor seconds primarily). These faded in and out unpredictably avoiding any sense of rhythmic synchronisation with Layer 1. Layer 3 consisted of variations on the Amelia vibraphone motif that were quite sparse including some silent gaps and again a very noticeable lack of obvious rhythmic pulse. Care was taken that any juxtaposition of this vibraphone motif with Layer 1 did not produce harmonic clashes of close intervals inadvertently. It was important to hold back the level of dissonance and tension for the more active and horrific sections of game play to follow. This in itself is interesting because it shows that although locally the structure of the game was non-linear, in a more macro sense there was still a degree of predictability in the order of events that could be used to structure the dynamics of the soundtrack. This turned out to be important in creating a sense of immersive narrative that would pull the game player onward through the game. For example the player might perhaps stumble across the mausoleum in the graveyard before acquiring the tools with which to unlock it but they could not enter until they had back tracked and retrieved the tools.

Layered on top of these three none synchronised musical loops were short sounds designed to be cued from in game events such as finding a key for the mausoleum. For these I used a collection of broken out of tune autoharps scavenged from junk shops. Various short dramatic sweeps and dissonant clusters were recorded and these sound design elements sat nicely between the more overtly musical material and environmental sound effects which were to be additionally layered on top. The autoharps were all used as found, completely out of tune! As a result they each had a wonderfully individual dissonantly spooky quality that provided just the right amount of subtle emotional jarring shock while still leaving room for further intensity in the later chase scenes. Because of the sporadic and unpredictable nature of the triggering of these autoharp events they further combined with the desynchronised loops to convey a convincing sense of continuously evolving musical score for this section. Importantly this evolving musical score would never be exactly the same two times in a row.

The final element in this soundtrack layer cake were the more traditional sound effects and foley that had been specified by the game developers. Footsteps on various materials triggered by the in game location of the player, rustling foliage, soft wind noise, night time ambience and so forth. These were again combined into two more desynchronised loops but this time the loops were more closely tied to the game payers actions. Here I was deliberately using sound produced by the protagonists body movements to try and extend the sense of self of the player into the game space. This was an additional measure to heighten the sense of emotional immersion and identification of the player with the game world. These loops consisted of rustling foliage triggered by exploring specific parts of the graveyard, overlapping with wind noise for more static situations in which the game player may just be looking around the environment. The wind noise was used as a particularly useful bridge between the more electronic aspects of the music and the sound effect layer. Some alterations were made in the electronic music sounds to blur the line between music and foley. This was a deliberate ploy to make all of the non dialogue soundtrack elements blend into a seamless whole that would further enhance player immersion. As work progressed it became clear that this seamless immersion was particularly important in eliciting the strongest emotional identification from the game player. I have described these various musical elements as loops or layers but Grey has made an interesting analogy with the use of modular concepts of composition emerging from the classical avant garde in the twentieth Century (Grey 2016). However Grey doesn't mention another interesting link with contemporary computer mediated dance music which builds music from modular overlapping musical cells which might be even more pertinent in the games context. For examples of this kind of modularity at work see the recorded output of bands such as Orbital and Leftfield.

Taking a critical step back from the actual creative process at work and trying to pin down more concrete examples of how emotional effects were being created gave rise to a flood of specific examples, more varied and detailed than can be covered here. However here are some representative examples of how particular emotional effects were manifested in this particular game scene.

Sound from an unseen source off screen was used to generate unease and tension. In this first game play level the Lovecraftian horror represented almost exclusively by the sound of giant flapping wings was made to sound more or less distant, to pan across the stereo field and to get louder and quieter. Some more subtle elements such as low pass filtering and psychoacoustic 3D spatial processing were also employed to create the effect of an approaching or receding unseen threat.

Music and sound design elements were used to create other types of anticipation as well as cueing various aspects of the game play. When the player approaches within a certain distance of the mausoleum, the music subtly changes to something more processional with dark religious undertones (Slowly tolling bell with a vocal chant reminiscent of Gregorian chant). The closer the player gets to the mausoleum the louder this music becomes therefore directing the player's attention to this area even before the mausoleum actually becomes visible. The music conveys a sense of ritual and trepidation as the player approaches. This transition to different processional music is handled by a slow crossfade in order to not break the seamless musical flow. Aspects of the atmospheric sound effects such as the wind noise continue across this musical transition and help maintain a seamless immersive continuity.

Once inside the mausoleum the player solves a puzzle to unleash the monster from a sarcophagus. At the point that the monster is unleashed there is cued a sudden abrupt and very jarring musical lurch to heavily dissonant screeching violins and brass accompanied at times by a demonically distorted reference to ragtime piano. The ragtime references historical styles of the 1930s but with added dissonance and pounding, heavily percussive playing. After a short while insistent high energy percussion something like distorted voodoo drumming is also added to the cacophony which gives a sense of fast paced excitement as the monster chases the main character out of the mausoleum and across the graveyard. To put it simply the energetic soundtrack invites an energetic gameplay response from the player who is galvanised into running away. This sudden jarring departure from seamless music is all the more effective at raising the emotional temperature as it comes after quite a long sequence of dark ambient style seamless musical underscore. By now the player has become so immersed in the atmosphere that this sudden shocking jolt can be employed to generate a lot of adrenalin and excitement without risking the player dropping out of their immersive state.

Music was also required to represent the capture of Amelia by the winged monster who could, depending on the game play, swoop down and carry her off up into the air. This was achieved by taking the frenetic chase music and electronically pitch shifting it slowly upwards. This gave two primary effects, the rising pitch represented the rise into the air without having to show the monster itself but perhaps more interestingly the pitch shifting effect squashes the normal distance between musical notes giving rise to micro intervals more normally associated with the outer limits of the classical avant garde. Subjectively this elicits the emotional response of the normal rules of reality (harmony) being distorted—a useful device for representing a breakthrough of a dark occult force into the gamer's normal reality!

Finally a rapidly descending whoosh followed by a gruesome 'splat' was created by similar pitch shifting means to denote Amelia being dropped from a great height—end of game level! This short and rapid descent effect was to be cued by a particular game action and superimposed over the queasy pitch bending cacophony of the aerial snatch scene. The screeching classical instruments, the voodoo drums and the demonic ragtime were all on independently looped layers of differing lengths so this texture could also be seamlessly extended according to the needs of the game play.

One final element of the sound design proved important in creating an effective sense of emotional immersion. This was the use of convolution reverb on various sound effects and musical beds to place them in a convincing three dimensional environment. Convolution reverb allows the accurate recreation of convincing acoustic spaces and proved particularly effective in defining the entry into the claustrophobic echoey vault of the mausoleum from the windswept exterior acoustic of the graveyard. At the time of writing work was still underway to define ways of controlling the relative amounts of this type of reverb within the game engine to denote the proximity of various game play assets.

Useability Functions

Another way of looking at the creation of effective emotional responses for the player can throw further light on the techniques utilised. These can be understood as a set of game play functions executed by aspects of the audio.

The spot cueing of the detuned autoharp gestures form a kind of 'earcon' the audio equivalent of a visual icon for a particular event. These are game specific and the significance of such earcons needs to be learned or rather recognised by the player. Once recognised the player feels a sense of satisfaction that they have successfully unraveled an unspoken aspect of the game structure. Similar to the player reinforcement provided by successfully solving a more overt puzzle feature. Atmospheric functions are those that enhance player immersion, world creation, fantasy, and create a sense of presence in the game world. There are also directing functions where sound is used for recognition, or prioritising attention. Here audio becomes an important part of the game mechanics. Taken together all of the above interact and form building blocks of a convincing emotional experience.

> Emotional cues in games such as music interact at a complex level with psychological aspects the game player brings to the game playing situation. Such as motivation and intent. (Swink 2009)

It might further be suggested that effectively realised audio useability functions go at least part way to creating or influencing these psychological states in the player such as the motivation and intent mentioned by Swink.

Further Developments

Having outlined at least some of the subjective techniques employed while working on *Deal with the Devil* it is tempting to look into the near future of immersive gaming. Judging by the extensive development work currently going on in the field the introduction of more widely available virtual reality or augmented reality gaming will almost certainly change the ground rules once again. Putting the player into a virtual environment calls into question many of the conventions of audio soundtrack creation. In particular what should be the 'point of view' of the audio in such a new context? In *Deal with the Devil* the player's visual viewpoint is predominantly third person trailing (Over the shoulder of the main character) with occasional use of third person omniscient for cut away shots. The audio has its own perspective though and is more concerned with immersion than accurately reflecting the position of the implied camera. Is this still appropriate for a virtual or augmented environment or will it work against player immersion? Is the use of headphones in the gaming context significant. Headphones tend to create the illusion of the sound coming from inside the player's head which can be relevant in some situations but not others. Will virtual or augmented reality change a fundamental bounding parameter for audio in a similar way to that of the introduction of non-linear form?

References

Baystead, S.: Palimpsest, pragmatism and the aesthetics of genre transformation. In: Kamp, M.S., Sweeney, T., M. (eds.) Ludomusicology: Approaches to Video Game Music, pp. 152–169. Equinox, Sheffield (2016)

Bordwell, D., Thompon, K.: Film Art, 5th edn, p. 315. McGraw Hill, New York (1979)

Collins, K.: Game Sound: An Introduction to the History and Theory of Video game Music and Sound Design, p. 3. MIT Press, Cambridge (2008)

Collins, K.: Playing with Sound: A Theory of Interacting with Sound and Music in Video Games, p. 9. MIT Press, Cambridge (2013)

Gonzalez, M., Haselager, W.F.G.: Creativity and self-organization: contributions from cognitive science and semiotics. SEED. **3**(3), 61–70 (2003)

Grey, E.M.: Modularity in video game music. In: Kamp, M., Summers, T., Sweeney, M. (eds.) Ludomusicology: Approaches to Video Game Music, pp. 53–72. Equinox, Sheffield (2016)

Jorgenson, K.: A Comprehensive Study of Sound in Computer Games, p. 179. Edwin Mellen Press, New York (2009)

Round Table Games, Deal With The Devil, trailer, Retrieved 17 May 2017, from Youtube https://www.youtube.com/watch?v=7QFn31tyeps

Sega: Rez. Tokyo, Sega (2001)

Swink, S.: Game Feel, p. 66. Morgan Kaufmann, New York (2009)

Van Elferen, I.: Analyzing game musical immersion: the ALI model. In: Kamp, M., Summers, T., Sweeney, M. (eds.) Ludomusicology: Approaches to Video Game Music, pp. 32–52. Equinox, Sheffield (2016)

Chapter 6
Brain Computer Music Interfacing (BCMI)

Duncan Williams

Introduction

This chapter describes the applications for systems which measure and respond to emotions as they manifest in the human brain. The most likely end use in a video-game scenario would, as you might expect having read previous chapters addressing some background to this topic, be emotionally-congruent sound-tracking. In the future, the system could be adapted to musical structures input by users in order to foster opportunities for further musical creativity; facilitating music-specific game-play. The long-term motivation is to allow non-linear modification of musical feature sequences in response to real time affective interaction, measured through autonomic means (specifically, biophysiological measurement).

The chapter will first introduce the term *Brain Computer Music Interfacing* (BCMI) before illustrating the use of BCMI techniques to estimate specific emotions. Finally, a combined BCMI-AAC system will be illustrated (see Chap. 4 for a generic introduction to AAC). This work builds upon work previously presented at the European Society for Cognition of Music (ESCOM) meeting in Manchester, UK, in 2015 (Williams et al. 2015).

Defining BCMI

The expression Brain–Computer Music Interfacing, or BCMI, was coined by Plymouth University's Interdisciplinary Centre for Computer Music Research team to denote BCI systems for Musical applications, and it has since been generally

D. Williams (✉)
Digital Creativity Labs, University of York, York, UK
e-mail: duncan.williams@york.ac.uk

D. Williams, N. Lee (eds.), *Emotion in Video Game Soundtracking*, International Series on Computer Entertainment and Media Technology,
https://doi.org/10.1007/978-3-319-72272-6_6

adopted by the research community (Miranda and Castet 2014). Research into BCMI involves three major challenges: the extraction of meaningful control information from signals emanating from the brain, the design of generative music techniques that respond to such information, and the definition of ways in which such technology can be deployed effectively; e.g., to improve the lives of people with special needs, address therapeutic applications, or artistic purposes. BCMI is a growing field, with a history of experimental applications derived from the cutting-edge of BCI research as adapted to music making and performance.

As we have already considered elsewhere in this book, music has often been considered the language of emotion—Music has been shown to be a useful vector for emotional communication, in that there are emotional responses that are reliably elicited by particular pieces, but the ability to compose, perform, or improvise in response to, or reflection of these emotions often requires a good deal of musical training (Lin and Cheng 2012) and shares two fundamental properties with BCI, generally, communication and interaction. Music has been shown to be capable of emotional contagion (Arizmendi 2011; Egermann and McAdams 2013) and to induce physical responses on a conscious and unconscious level (Anshel and Marisi 1978; Grewe et al. 2005; Grewe et al. 2007). Music can facilitate communication from or to the player, and even foster interaction between a gamer and other player characters. Listeners do not need any special musical education to understand communication made by musical means (Bigand and Poulin-Charronnat 2006; Bailes and Dean 2009). BCI offers the possibility of directly translating gaming actions to performance in soundtracking.

System Overview and Example

Affectively-driven algorithmic composition (AAC) combines computer-aided algorithmic composition techniques with cognitive approaches for the measurement of emotional responses in the listener, as described fully in Chap. 4. The aim of such systems is to generate new music which can deliberately induce particular emotional responses, specifically reflective of the current state of gameplay. More than simply reflecting emotions, another important aspect of AAC is the desire to develop technology for building responsive and intelligent systems that can induce specific affective states through the generated music, automatically and adaptively. In the case of AAC, there is the potential to develop systems which respond not only to user reports (i.e. self-report of emotion, which can be unreliable if the participant is not completely clear on the distinction between emotion intended and emotion felt, but also to physiological indicators of emotional states. Emotional responses to music can be dichotomous in this regard. If the listener is already in a negative state of mind, they may find that musical stimuli that match this emotional state actually increase their own positivity, thus the induced emotional response could be completely different to the perceived emotion reported by the listener—simply put, many listeners who already feel ‘sad’ can find ‘sad-sounding’ music uplifting or enjoyable (Vuoskoski et al. 2012). In such cases the distinction between perceived and induced affective state is crucial: the affective state of the listener must actually change in response to the musical stimulus. The distinction between perceived and induced has been well-documented by music psychologists and musicologists, although the specifics are not always universally agreed-upon (see Chaps. 2 and 3). The utility of methods for addressing the induction of an emotional response becomes clear if physiological correlates of emotional response could be determined accurately and utilised as a control vector for specific music generation by means of AAC: A system for creating music autonomously by combination of physiological analysis and affective musical correlates as dependent meta variables might allow gamers to reflect their emotional state, or interact with their narrative and other non-player characters musically without any prior musical training. The fully realized system should be able to monitor the affective state of an individual, and induce specific affective states through music—adaptively—according to their individual physiological responses and the current state of gameplay. The generated music might use the induced affective response as measured by, for example, brain-imaging, to inform the selection and use of whichever features are necessary in the music generation algorithm in order to target the intended emotional state in an individual listener.

A live demonstration of such a system, using a Yamaha Disklavier (a self-playing piano) in MIDI enabled mode, can be seen at https://www.youtube.com/watch?v=wDm1iqfHkag (Fig. 6.1).

The generation algorithm uses a multilayer feed forward network (there are no cycles; the interaction is created by adjusting the weighting at each node in the hidden layers). For any given musical feature, the existing generation is moved towards the target value by linear interpolation towards the closest value in the training material, which then supersedes the current feature value in the generation. For example, according to the mapping, co-ordinates with higher emotional arousal generally include a faster tempo, larger pitch spread, and harder timbres (with piano voicing the timbre of the performance can be manipulated using dynamics markings, where perceptually harder sounds are achieved with louder performance dynamics), whilst co-ordinates with higher emotional valence generally utilize a major key and a spread of pitch values, which are comparatively higher than those of the rest of the generated pool. The hidden layers are trained according to the weighting of each value in the seed material (off line), and the sum of the new input material and the original seed material at the activation function when the network is deployed online, such that the weighting of the hidden nodes can be adjusted from the initial values according to a real-time input performed by the user. Essentially, the weights become the memory of the neural network, with this memory being further influenced by any real-time input. This approach has been used for many supervised learning applications. There is a danger that this architecture would ultimately modify the network such that it played 'in unison' with the human, which might remove the possibility for affective steering of the players emotional state. This might be a fruitful avenue for further study. However, in many video game soundtracking applications, emotional mirroring in the performer is a characteristic

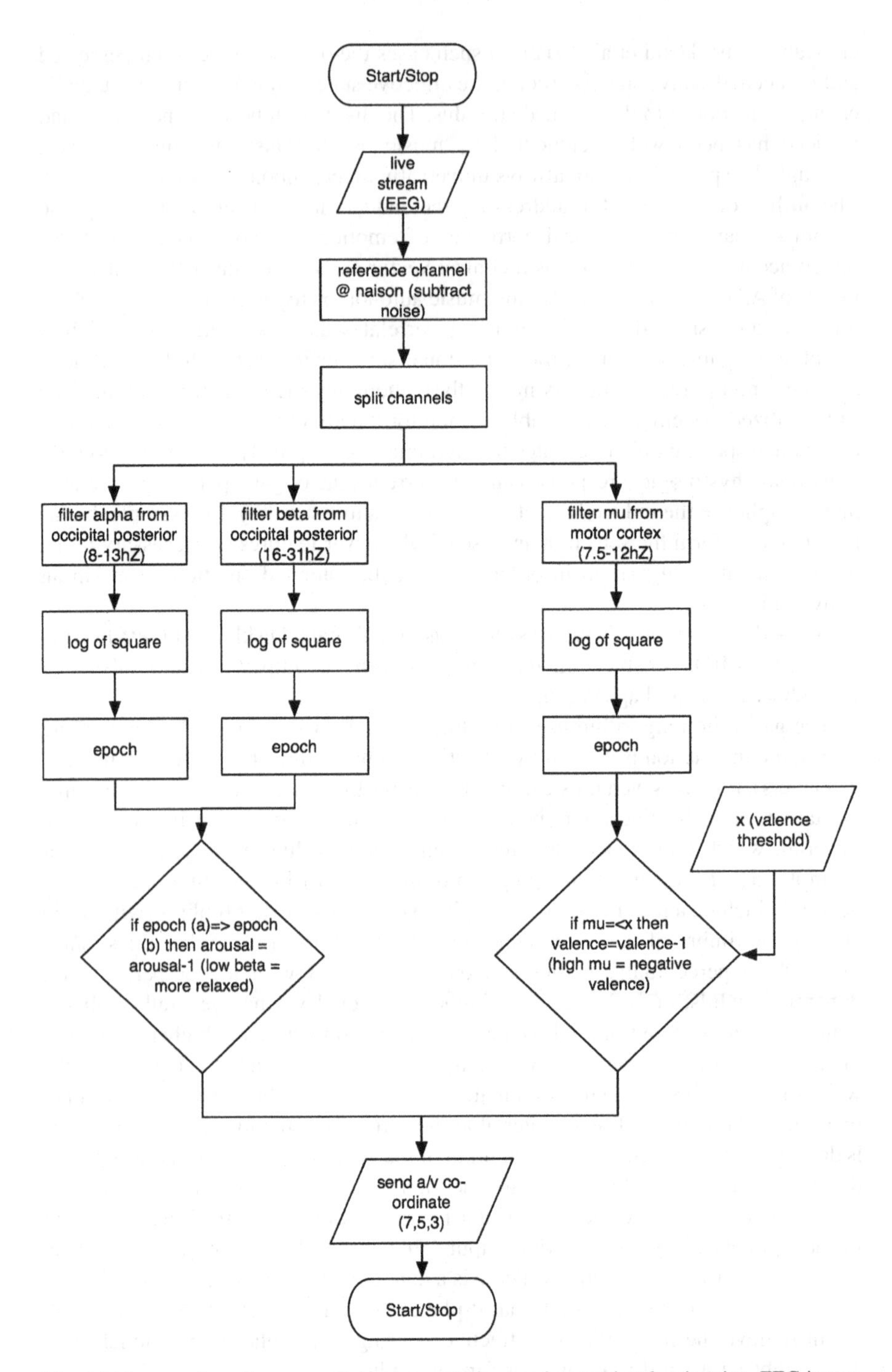

Fig. 6.1 An overview of one use of electroencephalogram for emotional estimation. EEG is processed into alpha, beta, and mu frequencies before being subjected to windowing analysis. Signals are then used to infer emotional characteristics in arousal and valence

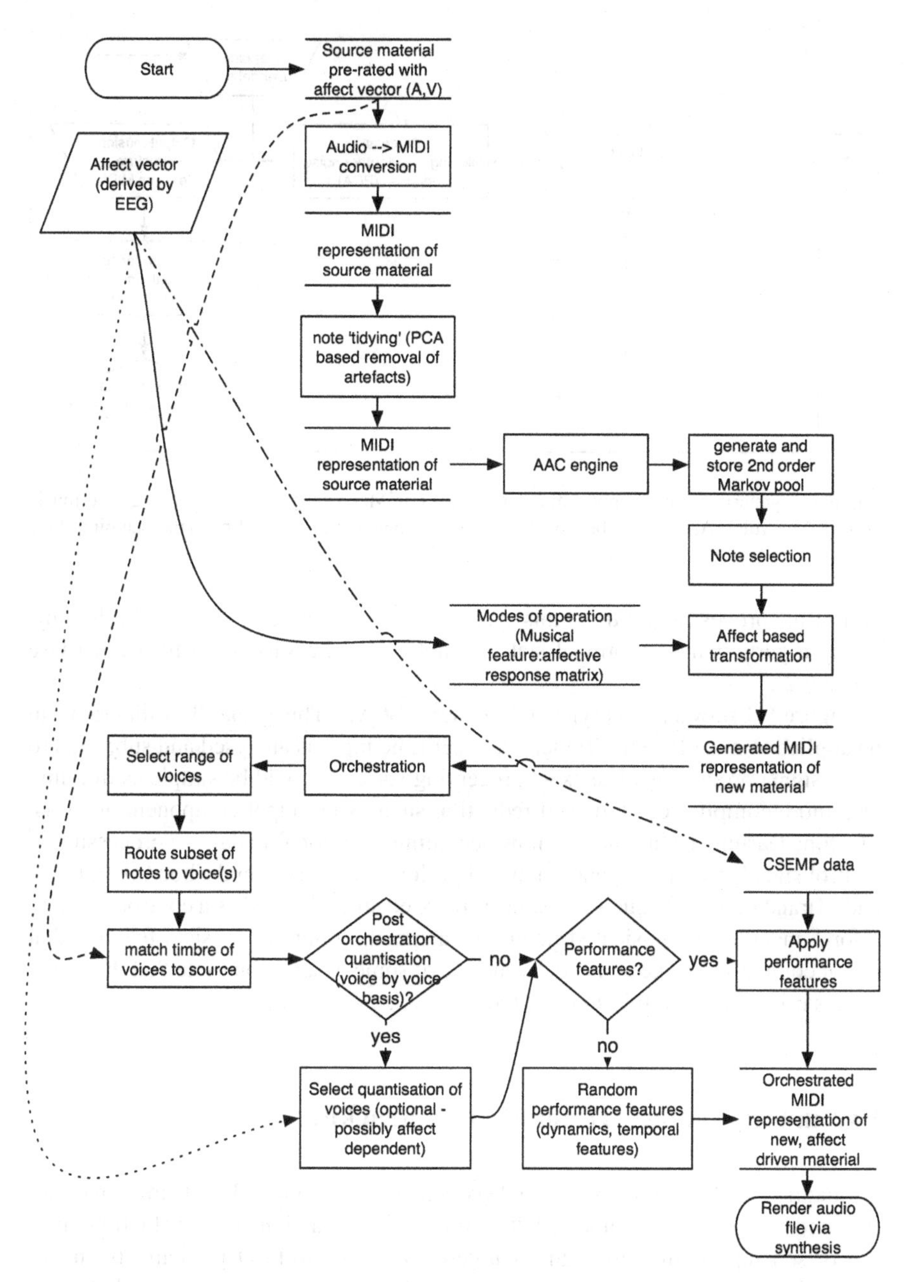

Fig. 6.2 Overview of the supervised learning algorithm adapted to real-time control of weightings for affectively-driven music generation in a video game soundtracking configuration

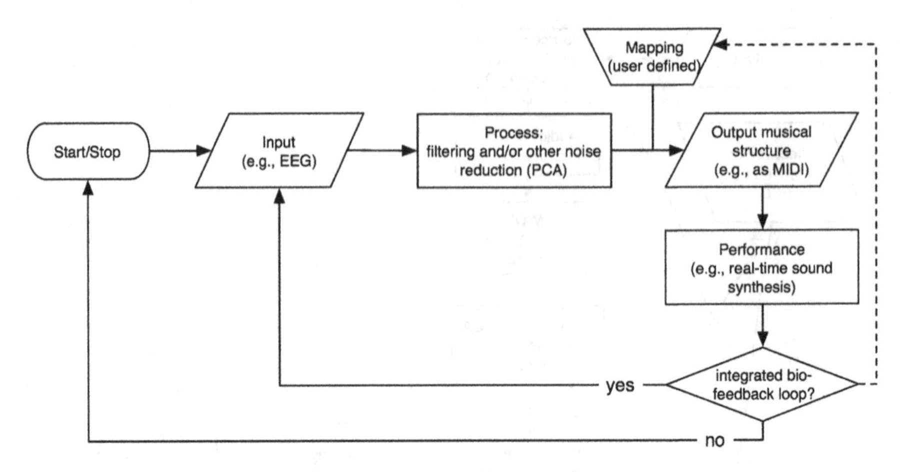

Fig. 6.3 A generic brain-computer music interface (BCMI), using filtering, mapping, and music performance rules. An optional bio-feedback loop is shown in the dashed line to the mapping stage

part of the process and thus, if this system was able to reach the level of fully mirroring a participants emotional state it might be more successful than otherwise (Figs. 6.2).

Figure 6.3 shows an overview of a generic BCMI. This signal flow diagram can be applied to most BCMIs. Typically, a real-time input is analysed and subjected to some signal processing. The exact processing varies, it could be simple as filtering, or a more complicated statistical reduction such as principal component analysis. Machine learning techniques are now becoming common for adaptive processing of control signals for music generation and performance (Jiang and Zhou 2010; Kirke and Miranda 2011). In such cases, the processed signal is used as a control signal to inform mapping to musical structure, or a specific range of musical features that might combine to make a musical structure of some description (note that this is not necessarily 'music' at this stage of the process).

Evaluating Against 'Human' Soundtracking

As mentioned above, the distinction between perceived and induced emotion (emotion understood, versus emotion felt) is crucial in evaluation. This evaluation made use of self-report, and although great care was taken to brief participants on this distinction, consideration of the results should be tempered by an acknowledgment of the difficulty in using self-reports when reporting on induced emotions, for which the use of biophysiological measurements in future might provide a useful circumnavigation. In this evaluation a range of stimuli covering the maximum possible subset of the 2-D affective space was drawn from (Eerola and Vuoskoski 2010) a dataset of excerpts from film scores spanning a range of timbres and styles rated

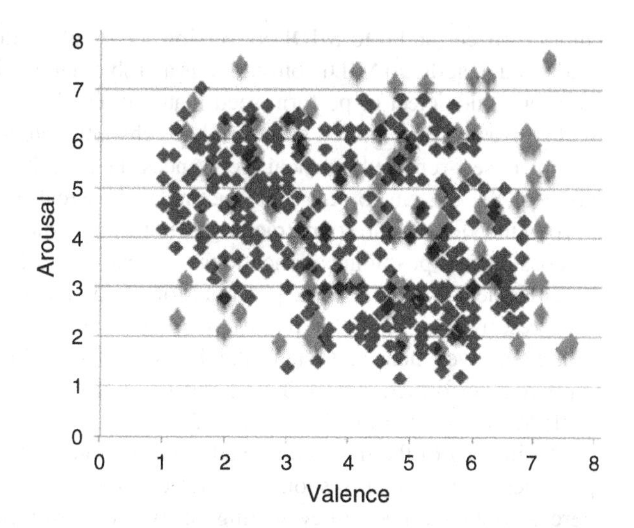

Fig. 6.4 Valence-arousal space showing listener responses to a stimulus set generated by AAC system (blue) after the affective mapping (relative contribution of specific musical features) was adjusted in a tri-stage perceptual experiment, in comparison to real-world stimuli (red)

according to valence and energy (where energy can be taken as synonymous with arousal). Stimulus sets were also generated by the prototype system and rated by a listening panel of 36 participants in a listening experiment. Their responses were then compared to the previously rated film music set. The resulting span of emotional responses demonstrated that the AAC system was capable of producing music with a full range of emotional responses, once the affective mapping has been appropriately adjusted according to the listener responses in the first two stages of the experiment. A summary of the resulting range of emotional responses across two dimensions is shown in Fig. 6.4.

Discussion

The evaluation of the system outlined above showed that this approach to AAC was capable of generating a full range of induced emotional responses in a listening panel. However, many of the individual responses exhibited significant variance from the average. The nature of individual emotional response to music suggested that a generalized system, particularly one with the intention of therapeutic use, was likely to be less than optimal. One longer term solution would include training machine learning algorithms to manipulate the specific contribution of the various musical features in the affective mapping, perhaps in response to biophysiological measures as mentioned above. However, there also remained the potential to explore adaptation of this system for creative purposes by enabling the user to select their own choices of seed material. Two possible approaches were implemented for evaluation. First, with the end use of a creative composition augmentation tool in mind, the system was adapted to allow the user to input their own musical structures in MIDI format for each of the nine points on the 3 × 3 grid across the two-dimensional emotion space (audio to MIDI conversion would also allow other file types to be used, though audio to MIDI conversion is a problematic field in its own right). Thus, the user could select or perform seed material which they felt best expressed particular areas on the emotion space, and thereby generate new material according to their own (self-reported) musical preferences. This might also facilitate future adaptation of the system to real-time dueting with a real-time game narrative input. Beyond the video game soundtracking opportunities such a system might offer, and the possible compositional applications for creative expression of emotion in music-driven games, there is also the potential for some practical and pedagogic benefits: some research suggests that pianists perform duets better when they are duetting with themselves, hence a system like the one demonstrated above could be used to improve a performers own skill as a musical training tool (Haueisen and Knösche 2001; Meister et al. 2004; Keller et al. 2007).

The infancy of the field, combined with previous difficulty in ascribing specific quantities of emotional response to musical features (and vice versa) means that there is no standard way of evaluating this type of system, a dilemma which is exacerbated if the creative output is to be objectively assessed, as in the case with the user supplying their own seed material to train the generator for the various areas of the emotion space. However, without some method for evaluating the output of this type of system, a fully realized system for monitoring and inducing affected states, as outlined in the introduction to this chapter, becomes increasingly problematic. Therefore, we propose and attempt the adaptation of a methodology for evaluating creatively generated music by means of multi-criteria decision-aid analysis, as is often found in the world of sonification (alternately referred to as auditory display), which is a growing field that has already addressed many of the challenges which face creative adaptation of existing data to sound and music (Ibrahim et al. 2011). We employ a similar approach, but with a revised criteria considering the specific circumstances of the AAC system and the three discrete approaches to training the generative algorithm that have been implemented to date. Sonification is often used to augment visual analysis and therefore typical evaluation criteria can include reference to the utility of the auditory display, as well as the immersion with the data and the intuitivity of the sound in a multimodal context. These concerns are less relevant to an AAC system, but criteria such as amenity (how enjoyable or aesthetically pleasing), and congruency (particularly, in this case, how emotionally congruent the music was) may both have practical implications in the evaluation of these types of system. Creativity, as a third criterion, may be relevant to some types of auditory display, though it is likely more relevant to systems for music generation than any other type of system (it would be irrelevant to the evaluation of an audio alarm system, for example).

Three algorithmic variations (seeded with random neutral material, user input, and user ranked input) were evaluated by the listening panel. All had some experience of critical listening as part of undergraduate level study in music. The study was given ethical approval by the research ethics sub-committee of the Faculty of Arts, Plymouth University.

For each algorithm variation, participants listened to 18 pieces of generated music (two pieces for each discrete area of the emotion space derived by dividing the valence and arousal space into a 3 × 3 grid) and were asked to rate each piece on amenity, emotional congruence, and creativity as outlined above. The ratings were entered on 9-point Likert scale, labeled from 'least' to 'most', with high numbers indicating a good agreement and low numbers indicating a disagreement with the term being rated. The experiments were conducted via a browser based interface in real-time, whereby participants were asked to rate the three criteria across each algorithm 18 times (twice for each discrete area of the emotion space, as divided in 3 × 3 cartesian co-ordinates in valence and arousal). Playback was conducted via circumaural headphones in a quiet room.

Results

The participant ratings for the criteria (amenity, emotional congruence, and creativity) across each of the three algorithmic iterations are shown in Tables 6.1, 6.2, and 6.3.

Table 6.1 suggests that participants found the user-rank seeded generations the poorest in terms of amenity, and the user-input seeded generations the best in terms of amenity. Without fully quantifying the difference between a generation and the seed material it is difficult to say that this might not be because the user-seeded generations were simply preferred by the listeners, particularly as the neutrally-seeded generations were rated roughly in-between, but with more disagreement shown in the standard deviation. Again, this highlights some of the methodological difficulties inherent in conducting evaluations of this type. The p-value for all ratings was <0.05. Table 6.2 suggests that participants found the neutrally-seeded generations the least emotionally congruent, with the user-rank seeded generations most emotionally congruent, and the user-input seeded generations falling roughly in-between. This provides some balance to the possibility of the generated material being simply very similar to the user input generations. However, standard deviation was slightly higher for two of these rankings when compared to amenity, in particular for the neutrally-seeded generation, which also exhibited a p-value above the level of statistical significance. Table 6.3 suggests that participants found all of the generations approximately comparable in terms of creativity, with the user-input seeded generations being marginally more creative (and exhibiting lower standard deviation than the other two rankings). However, the standard deviation for all of the rankings for creativity was higher than those for amenity or emotional congruence, and the p-value for all of these ratings was above the level of statistical significance, >0.05. Beyond suggesting that there was little agreement in which was the most creative solution, this might also highlight (and confirm) the more general difficulty in objectively evaluating creativity in computer music.

Table 6.1 Likert-scale responses (mean, standard deviation, p-value[u5]) showing participant ratings for amenity to algorithmic generations from neutral seed material (the original prototype system), user selected seed material, and user ranked seed material

Seed type	Mean	Std.D (σ)	p-value
Neutral material	5.1	1.5	0.031
User input	3.5	1.0	0.028
User ranked	6.3	1.3	0.041

Table 6.2 Likert-scale responses (mean, standard deviation, p-value) showing participant ratings for emotional congruence to algorithmic generations from neutral seed material (the original prototype system), user selected seed material, and user ranked seed material

Seed type	Mean	Std.D (σ)	p-value
Neutral material	2.3	2.5	0.061
User input	6.8	0.9	0.035
User ranked	4.8	2.1	0.029

Note that the p-value of the neutrally seeded generation ranking is above the level of statistical significance, <0.05 (highlighted in bold)

Table 6.3 Likert-scale responses (mean, standard deviation, p-value) showing participant ratings for creativity to algorithmic generations from neutral seed material (the original prototype system), user selected seed material, and user ranked seed material

Seed type	Mean	Std.D (σ)	p-value
Neutral material	5.3	2.3	0.122
User input	5.8	1.2	0.089
User ranked	5.5	3.3	0.101

Note that the p-value of all rankings was above the level of statistical significance, <0.05 (highlighted in bold)

So where does this leave us? Taken in combination, these results make it difficult to draw strong conclusions. It appears that there was only a marginal improvement (at least in terms of the criteria evaluated here) between the different types of seed material used by the AAC system. However, the highest rated type of seed material across all three criteria was the user-input seed, with rankings that were $p = >0.05$ in two out of three criteria: amenity and emotional congruence. Assuming that creativity is a difficult criteria to evaluate, these results suggest that some improvement in the performance of an AAC system can be achieved by allowing individual selection of training material. This suggests that individually calibrated AAC systems would be a fertile avenue for further work.

Concluding Remarks

This chapter describes an AAC system for generating music using an artificial neural network algorithm, fed an emotional target by a brain–computer interface. The AAC system generates music according to an affective mapping of musical features with emotional correlates, adapted towards individualized music generation by enabling users to either input their own material to train different areas of a two-dimensional emotion space, or to rank existing material with which the areas of the space would be calibrated before generating new music. The basic technique for generating music described here is not novel in and of itself. Many other successful implementations of different types of probabilistic analysis and subsequent music generation exist, including heuristic Markov models and recurrent neural networks, amongst others. However, the combination of generative AI techniques with affective feature mapping does offer a novel way of creating and interacting with emotionally expressive music, with enormous potential in the context of creating opportunities for musical communication, particularly with a virtual player as is the case in video game soundtracking. The field of AAC is in its infancy and therefore there are no major precedents for evaluation of such a system. Methods for evaluating the utility of music creation systems, particularly algorithmic composition are the subject of much debate, including debate as to the question of authorship of music created by such systems. This system was subjected to a multi criteria decision aid analysis, inspired by the approaches to evaluation that recent research in auditory display has utilized. Three criteria were adapted for the evaluation: amenity, emotional congruence, and creativity. A listening panel was invited to rate material generated by the various seed types ('neutral' material, personally input material, and personally ranked material) using a 9-point Likert-scale. Statistical analysis of the responses suggested that some improvement to amenity and emotional congruence could be achieved when the system was trained with individually input material. Responses to the criteria 'creativity' suggested that there was a lack of consensus amongst participants as to whether any improvement was achieved, or perhaps what creativity was, in and of itself. This might come as no surprise given that the evaluation of creativity in computer-aided composition is a complex field.

Practical improvements might refine approaches to the number of higher-level musical features, particularly those with well-documented emotional correlates. The complex nature of the inter-relationship between a full range of musical features, and the subsequent affective responses that might be the correlations thereof, remains an area of considerable further work. Expectation and surprise, for example, are temporal musical features which could not be readily mapped using the small number of features implemented in the current system, yet they are both thought to exhibit strong emotional responses in listeners. This could be further extended to overall musical structure, which is a hierarchical property that has been shown to have a huge influence on the perceived emotional characteristic of music (Gabrielsson and Lindström 2001). Structure remains extremely challenging for any algorithmic composition system, and AAC systems, such as the one presented here, are no exception, particularly when being designed for use with video game soundtracking. Finally, in future work, the potential to harness brain–computer interfaces in the control of an AAC system provides an opportunity to respond individually to the emotional outcome of a musical feature set according to brain activity, rather than using the absolute mapping between the musical features and emotions evaluated here (making the mapping dynamic, and relative to the ongoing evolution of a user's emotional state). Thus, this type of system could be adaptable to real-time control by physiological sensing of affective states in order to create a feedback-driven system for simultaneous monitoring and induction of targeted affective states in the user according to specific gameplay narratives. There remains significant potential to derive a robust paradigm for combined affective evaluation, and to use this to explore other musicological correlations for specific affect (for example, the impact of timbre, surprise, expectation, and polyphonic generation). The challenges are many, but the possible rewards from the applications for this kind of combined system are numerous. Such a system would open exciting new doors for both creative music production and therapeutic tools for both emotional entrainment and emotional communication via music.

References

Anshel, M.H., Marisi, D.Q.: Effect of music and rhythm on physical performance. Res. Q. **49**, 109–113 (1978)

Arizmendi, T.G.: Linking mechanisms: emotional contagion, empathy, and imagery. Psychoanal. Psychol. **28**, 405 (2011)

Bailes, F., Dean, R.T.: Listeners discern affective variation in computer-generated musical sounds. Perception. **38**, 1386–1404 (2009). https://doi.org/10.1068/p6063

Bigand, E., Poulin-Charronnat, B.: Are we "experienced listeners"? A review of the musical capacities that do not depend on formal musical training. Cognition. **100**, 100–130 (2006)

Eerola, T., Vuoskoski, J.K.: A comparison of the discrete and dimensional models of emotion in music. Psychol. Music. **39**, 18–49 (2010). https://doi.org/10.1177/0305735610362821

Egermann, H., McAdams, S.: Empathy and emotional contagion as a link between recognized and felt emotions in music listening. Music Percept. **31**, 139–156 (2013)

Gabrielsson, A., Lindström, E.: The influence of musical structure on emotional expression. In: Juslin, P.N., Sloboda, J.A. (eds.) Music and Emotion: Theory and Research Series in Affective Science, pp. 223–248. Oxford University Press, New York, NY (2001)

Grewe, O., Nagel, F., Kopiez, R., Altenmüller, E.: How does music arouse "chills"? Ann. N. Y. Acad. Sci. **1060**, 446–449 (2005)

Grewe, O., Nagel, F., Kopiez, R., Altenmüller, E.: Listening to music as a re-creative process: physiological, psychological, and psychoacoustical correlates of chills and strong emotions. Music. Percept. **24**, 297–314 (2007)

Haueisen, J., Knösche, T.R.: Involuntary motor activity in pianists evoked by music perception. J. Cogn. Neurosci. **13**, 786–792 (2001)

Ibrahim, A.A.A., Yassin, F.M., Sura, S., MacDonell Andrias, R.: Overview of design issues and evaluation of sonification applications. In: IEEE, pp. 77–82 (2011) https://doi.org/10.1109/iUSEr.2011.6150541

Jiang, M., Zhou, C.: Automated composition system based on GA. In: 2010 International Conference on Intelligent Systems and Knowledge Engineering (ISKE), pp. 380–383 (2010)

Keller, P.E., Knoblich, G., Repp, B.H.: Pianists duet better when they play with themselves: on the possible role of action simulation in synchronization. Conscious. Cogn. **16**, 102–111 (2007)

Kirke, A., Miranda, E.: Emergent construction of melodic pitch and hierarchy through agents communicating emotion without melodic intelligence. In: Proceedings of 2011 International Computer Music Conference (ICMC 2011). ICMA (2011)

Lin, C.-Y., Cheng, S.: Multi-theme analysis of music emotion similarity for jukebox application. In: 2012 International Conference on Audio, Language and Image Processing (ICALIP), pp. 241–246. IEEE (2012)

Meister, I.G., Krings, T., Foltys, H., Boroojerdi, B., Müller, M., Töpper, R., Thron, A.: Playing piano in the mind—an fMRI study on music imagery and performance in pianists. Cogn. Brain Res. **19**, 219–228 (2004)

Miranda, E.R., Castet, J. (eds.): Guide to Brain-Computer Music Interfacing. Springer, London (2014)

Vuoskoski, J.K., Thompson, W.F., McIlwain, D., Eerola, T.: Who enjoys listening to sad music and why? Music. Percept. **29**, 311–317 (2012)

Williams, D., Kirke, A., Miranda, E., Daly, I., Hallowell, J., Hwang, F., Malik, A., Roesch, E., Weaver, J., Nasuto, S.: Affective calibration of a computer aided composition system by listener evaluation. In: Proceedings of the Ninth Triennial Conference of the European Society for the Cognitive Sciences of Music (ESCOM2015) (2015)

Chapter 7
When the Soundtrack Is the Game: From Audio-Games to Gaming the Music

Alexis Kirke

Introduction

In most games, sound plays second fiddle to the visuals. There is always much talk about the graphical rendering power of a computer or how beautiful and/or smooth the visuals are in a new game. There are of course games that combine sight and sound but where sound plays a larger part than normal—for example Guitar Hero, in which the user attempts to play along to popular music with a fake guitar. However, these could still not be described as "audiogames", as without the visuals they are unusable. The emergence of the iPhone and other smartphones has led to a number of innovative mainstreams games being produced which are sound only, such as the Papa Sangre series. In fact audio-only games for people with sight difficulties has a long history going back to the 80s.

This chapter introduces the concept of a game—Musicraft—that is both audio-only, and in which the soundtrack is the game, and which can be played as a game or an amateur composition tool. Musicraft draws most of its inspiration from the popular children's game Minecraft (Short 2012), a game which cannot be played by audio-only. Minecraft has two modes of play: Survival and Creative. Survival is the default mode and involves harvesting resources to build structures and protect one-self from roaming monsters, and to survive and become more powerful. For most players Survival play is an enjoyable and thrilling experience. As a result they learn the ins and outs of the building and harvesting interface without any effort—they *want* to beat the game. Later many players switch to Creative mode. In this mode, the player cannot be harmed. They focus on using the skills they learned in Survival mode to contribute to the environment, often on shared servers so they can show off

A. Kirke (✉)
Interdisciplinary Centre for Computer Music Research (ICCMR), Plymouth University, Plymouth, UK
e-mail: alexis.kirke@plymouth.ac.uk

D. Williams, N. Lee (eds.), *Emotion in Video Game Soundtracking*, International Series on Computer Entertainment and Media Technology,
https://doi.org/10.1007/978-3-319-72272-6_7

their creations to their friends. The game becomes a creative and social experience, as players build structures (e.g. castles and cities) and devices (e.g. mechanical contraptions) and are able to enter and interact with them. The audiogame Musicraft is conceptualised to play as a game, during which the players learn how to construct music in an abstract audio-only world isomorphic to elements of common music notation. The music constructions are focused on surviving and avoiding musical monsters. Later they can move into a creative mode where they simply play the game to create music they like. A secondary inspiration for Musicraft is the audio memory games like Simon and Touch Me, where the user must recall the order and location of certain pitches to win the game. As will be seen, by its very nature as a truly musical game, Musicraft is a non-persistent looping world that requires a memory model to be built in the mind of the player.

Overview of Past Audiogames

Audio games as a concept have been around for many years. There have been a number of useful overviews published (Mangiron and Zhang 2016; Rovithis et al. 2014; Yuan et al. 2011; Westin et al. 2011). The first reported audiogame sold was Atari's Touch Me game (Beksa et al. 2016) in 1974. This was a memory game where a series of tones would play, and user tried to recall the tone sequence and press appropriate buttons to mimic it. Although it was also a visual game (lights flashed in concert with the tones), the game was simple enough that it could be played by people as a pure audiogame. A similar handheld game was released called Simon (Rovithis 2012) in 1978 which was even more popular. In the 1980s a number interactive fiction or adventure games were being developed. These involved users being given text descriptions of a location ("you are in a large cave" or "you are on the bridge of a spaceship", etc.) The user could then enter text to say what action they wanted to do ("go north", "pick up gun", etc.) The gameplay normally involved a number of puzzles presented in descriptions and solved by entering a series of commands. By connecting these games to a text-to-speech interface, they could be made playable by sound only (Spiel et al. 2014).

Real Sound—Kaze No Regret was a game released for the Sega Saturn and Dreamcast consoles in 1997 and 1999 respectively (Fizek et al. 2015). Like many audiogames, it was designed for people who are blind or partially sighted. It can be thought of as an interactive radio drama. The user listens to segments of story and the then game chimes and then can make selections which decide how the plot unfolds. Audio Space Invaders (McCrindle and Symons 2000) was a PC game that used 3D audio called ambisonics (Frank et al. 2015), designed for a four speaker system, but which can apparently work adequately on a two speaker set-up. The game play consists of shooting flying invading aliens. It does not require a graphical interface to be played but, like many audiogames, provides a graphical interface in addition. This is a directional audiogame—much of the game is based on the user hearing where the enemy is: to their left, right, in front, behind. Different sounds represent different types of enemy ships, the Doppler effect indicates a ship's direction of movement, and pitch represents closeness to the player.

The Blind Eye was released in 2000 by The Blind Eye Research project (Beksa et al. 2016). Once again the focus of the game was on binaural 3D sound rather than speech sound. Binaural sound enables the simulation of surround sound on a pair of headphones/earbuds (Rumsey 2012). The game is a timed search-and-collect task set in a city environment. The user must find a series of musical instruments hidden in locations whilst avoiding cars or colliding with walls. The city sound environment is simulated for the user. In addition the user's footstep sounds change as they walk on different types of surface.

Shades of Doom, released in 2001, is an audiogame in the spirit of Doom, a new first-person shooter built from the ground up (Parker and Heerema 2008). It uses surround sound, and the user can hear things like the sound of wind in passages, footsteps echoing, and nearby equipment. There is also an optional voice-based environment analyzer. AudioBattleShip (Sánchez et al. 2003) is also designed for people who are blind. Once again 3D sound is used to implement the game, in this case the traditional battleships board game. Players are can hear the direction of their bomb drop, together with the result. A tablet is used for input, and sonic cues can also be given to remind the user where they have placed their ships.

Drive is an audio-only "racing" game released in 2003 (Velleman et al. 2004). However rather than focusing on steering (like most racing games) it focuses on speed, which the player must maximise. The user drives their car over power-ups spread along the track which they pick up and activate with keypresses. As they speed up the power-ups get harder to pick up. The speed also causes a musical soundtrack to change tempo and a co-pilot's voice to change behaviour. There are various other sounds in the environment designed to distract the user from their task, such as doppler-effected sirens and helicopters. Another driving-related game was made by the company that made Shades of Doom: GMA Tank Commander (Denham and McComas 2004). It is based on the classic arcade game. This is another surround sound game. It includes audio targeting systems, the sounds of approaching tanks, environmental ambience and audible recommendations from a tank crew.

Terra formers (Westin 2004) is a sci-fi action and adventure game which uses 3D sound to provide a sonic compass, and sonar to provide a sense of distance from objects in the direction the player is turned. There is also voice guidance. Footstep sounds indicate what the user is walking on. The player has to walk through different rooms and levels solving problems, and avoiding robots and other danger, to progress through the story. As mentioned, early audiogames were often based on text adventures. The Last Crusade is a text-based role playing game (RPG) released in 2004 (Dwyer and VanLund 2008). The game speaks out text but also gives useable items their own name sound. They also have a unique sound when they used by the player. Similarly with mobs and non-player characters.

Some research has been done on extending non-audio games into the audio realm. An example—in the spirit of Shades of Doom—is AudioQuake (Atkinson et al. 2006), which does not build a game from the ground up, but aims to make the groundbreaking first-person shooter Quake playable without visuals. The player has certain extra objects, a radar scanner, and an item-detection scanner. These give audio outputs and alert the player to an object or event. A more obvious candidate for audiogame conversion are music games designed for sighted-people, such as dance games and games where the user plays a fake guitar to attempting to synchronize with the music. In "Finger Dance", a new game rather than an audio overlay, users listen to music, and try to match the rhythm with key press patterns using four keys, responding to different types of drum rolls. There is also an audio-based menu system. Another audiogame from the music/dance genre is AudiOdyssey (Glinert and Wyse 2007). The player is a DJ and uses the keyboard or Nintendo Wiimote to play dynamically generated sounds. The aim is to trigger them in time with the current song playing to excite the virtual audience on the dancefloor. Stereo spatial cues tell the user how to move the Wiimote.

Speed Sonic Across the Span (Oren 2007) is a platformer audiogame that uses voice, auditory icons and earcons. Objects are represented to the left and right of the player by using stereo panning. The player can jump, walk and stand still. Jumping causes the pitch to go up and then down and the player lands with a noise. A voice announces which platform level platform the player is on (1, 2, 3 or 4). Running or jumping into a solid surface causes vocal sounds to trigger. A form of sonar helps the player find platforms in front of them. The sounds of enemies (bees and dogs) are panned so the player can tell what direction an enemy is approaching from.

Blind Hero is a music rhythm game (Yuan and Folmer 2008)—but unlike the earlier Finger Dance, it is an overlay on the Frets on Fire, a game similar to the well-known guitar hero. Blind Hero involves a specially-designed haptic glove that the user wears. The glove gives vibrations on different fingers to indicate that there are notes arriving that need to be played. As well as engineering the glove, much work had to be put in to design the timing of the vibrations, which indicate both how long to press the fake guitar buttons and how long to press them.

The release of the Apple iPhone had a significant impact on audiogames as it created a powerful piece of technology that was heavily headphone based, and had a very small screen. The 2010 game Papa Sangre (Östblad et al. 2014) was an audio-only released for iPhone which won or was nominated for a number of gaming awards, and fascinated both people who were blind and people who were not. It is based in a supernatural world. The player moves by tapping the screen as a footstep metaphor, and turns by turning the phone. The game uses binaural audio to create a surround effect in the headphones. The player must avoid hazards, solve problems and collect musical notes in the environment. A sequel was also released. Sound Swallower (Beksa et al. 2015), another iPhone game, was published in 2011 and is an audio-only game that uses the device's GPS. The player must avoid the Sound Swallower and run to collect sounds from around them before they are erased.

The Nightjar (Mangiron and Zhang 2016) is an iPhone and uses the same engine as Papa Sangre. Rather than supernatural world it is based in a science fiction environment, on a space ship. It utilizes the voice of the ship's computer and a separate voice guide, sometimes they deliberately contradict each other. The game received much mainstream publicity partly because the guide voice was provided by the popular actor Benedict Cumberbatch. Blindside (Brieger 2013) was a post-apocalyptic themed adventure story game released in 2012. It is compatible with Windows and Mac as well as iOS. In the game the player wakes up to find that they, and most of the people in their city, have gone blind. The player must navigate the city streets avoiding monsters, and discover more about what has happened. On the iPhone screen touch allows the player to walk around and phone tilt controls other movement and direction.

Released in 2014, Audio Defence: Zombie Arena (Brammer 2016) was a more fast-paced game than some of the iPhone puzzle solvers. It is a 3D sound first person zombie shooter. The user can hear the zombies approaching around them. They must turn to the face the zombie and fire a gun to destroy them. The user can select different guns with different shooting sounds and abilities, including a sonic cannon.

Moving back to the adventure/puzzler type iPhone games, A Blind Legend (Csapó et al. 2015), is an audio only action adventure game. The touch screen is used for general movement, and for combat actions. There is also a companion who provides a voice guide, saying things like “Look out—behind you!” A related game, but developed for PC, is Legend of Iris (Allain et al. 2015). It is a game for children who are blind, but has a specific non-entertainment function. Its aim is to help train the children’s navigation skills. Although it is designed for audio only A Blind Legend uses Oculus VR to take advantage of the headset’s directional audio system. Thus players can move their head to change the sound direction. The adventure game involves a number of puzzles is inspired by Legend of Zelda and it’s goal is to get a giant out of a spirit village. Solving puzzles involves practicing skills such as locating the origin of a sound and then remembering that location, and following moving objects by sound only. Interestingly the game was designed in collaboration with a composer who lost his sight at the age of 14.

Looking over the last few decades of audiogame releases and research, some patterns can be discerned. There is often a direction-based component, and the worlds created are usually attempts to simulate (even abstractly) a real world. Another common method is the use of speech to build environments. Where music is used, it tends to be a background element, except in the music/dance games. It is interesting to note that games such as AudioQuake and AudioBattleShips were partly built to provide extra social activities for people who are blind, and that Legend of Iris aims on providing an educational tool. This is of relevance to the game we will describe in this chapter.

The system now discussed in this chapter Musicraft, can be compared to Legend of Iris in an educational sense, and to the music/dance games, in that the environment is music not a simulated real world. Unlike most music/dance audio games, in Musicraft the music is the environment—not part of the environment. The main addition of this type of game to the field is that it has a creative mode. The fact it provides an intuitive ability to learn common music notation and can be viewed as a form of memory game, are auxiliary additions, which would require significant extra testing.

Fig. 7.1 A standard Minecraft survival environment

Musicraft

The audiogame designs discussed in the previous section do not have a seriously creative musical component, even Guitar Hero uses a highly simplified guitar and musical notation. There are iPhone games like Isle of Tune on iOS (Consalvo 2011) that are quite sophisticated and creative toys for building tunes. However Isle of Tune is more of a toy and is graphically-based, not accessible as such.

Most audiogames that are based on a "geography map" the physical geometry into a sound-space experience across the headphones. Musicraft takes a different approach, mapping a geography across time rather than spatialization. The map is metaphorical rather than representative. It acts as though the elements in a bar of common music notation have physical properties, rather than mapping a physical space onto the bar. Also, like Minecraft, it is both a game and creative and can be played ignoring all of the game elements to make music of different genres.

The simplest way to explain Musicraft, is to introduce Minecraft. Minecraft is initially a survival game. To win the game one must collect resources in an environment to defend against mobs (the Minecraft name for monsters) and starvation. Fig. 7.1 shows a typical Minecraft survival environment. The entire environment is potentially collectable, rebuildable and restructurable. As a result Minecraft also became very popular in a mode known as Creative Mode. Fig. 7.2 shows an example result of creative mode. In creative mode, one cannot die, and the mobs cannot harm you. In fact you can generate mobs. A player can fly through the air and instantly get any building/resources they desire from a menu rather than having to seek them while avoiding mobs. Fig. 7.3 shows the most basic building block of Minecraft. It can be seen that much of Fig. 7.1 is made using this block. In survival mode the player has to build increasingly powerful tools and weapons to mine better

Fig. 7.2 A Minecraft environment made by a user in creative mode

Fig. 7.3 An example building block from Minecraft

Fig. 7.4 A basic Musicraft World

resources and kill more powerful monsters. But in creative mode these are all available instantly.

Most people start playing Minecraft in the survival mode, and then many go on to play it in creative mode. The players often build huge and complex structures and machines which could be shared with friends by logging onto the game on shared servers. Minecraft evolved into a giant dynamic shared Lego as well as a game.

Musicraft is built in a similar spirit. It is designed to first be played as a musical game where the player avoids mobs. However, the skills that are learned playing this game can eventually be used in a creative mode, in which the player can be immune to mobs and build their own dynamic musical worlds. In fact the mobs become a part of that musical world. The game also has the potential to implicitly aid the learning of common music notation, and to exercise audio-based cognitive skills due to its method of looping rather than the standard soundscape. The last of these elements needs a great deal more investigation, as such a looping approach may make the game less attractive, and there is only limited evidence for how games can improve cognitive function (Rebok et al. 2014).

The transition from survival to creative mode is the element designed to make Musicraft unique. Players should not feel they are being forced to become skilful at making music with common music notation, but at surviving a game. It is only afterwards they realise they have now become algorithmic music makers. That they have learned certain principles of music while playing the game. (Note that although Minecraft has block types which are musical, the use of such blocks and Minecraft in general is not accessible to those who are blind.)

So Musicraft is a highly simplified 2D Minecraft, whose "block types" are musical notes in common music notation twinned with a timbre. Movement paths are only available on lines or spaces in the staff. The player is a quarter note starting at, without loss of generality, middle E at the beginning of the looped bar. To go above middle E the player can "jump" (by, for example, pressing the space bar) but will return to middle E unless they jump in the direction that has a note that is low enough for them to climb onto.

Figure 7.4 shows a basic Musicraft world. Worlds can be of different lengths and tempos. They can also be of different tonalities and timbre sets. This is an 8 beat world and, we will say, 100 beats per minute world. A world is audio-displayed by playing it as a loop. The bass clef in the following example is not the "ground" but the backing. It is designed to provide the beat and also to make it clear where abouts in the display loop the output is. That is the reason for the first four note pitches in

Fig. 7.5 Player moves to the right in the world

the bass clef being different to the second four notes. The single note in the treble clef represents the player and their location. They are at the "leftmost" point in the 8 beat world. The player can move left, and right and jump.

This makes clear a key feature of Musicraft—it is non-persistent. Most previous audiogames create a sense of a persistent world. The whole world can be heard at the same time. In Musicraft the world is non-persistent and requires the user to listen to the whole loop to know what is in the world. Note that this does not mean that Musicraft is a sonification (Kramer 1993) of Minecraft in any sense. The Musicraft world is the common music notation.

Jumping will move the player up one note for one cycle. A cycle if one playing of the World. Then the next cycle the player will return to the lowest line. Moving left and right will move the player back in time or forward in time. If the player is at the start of the world (bar) they will wrap around to the end of the bar (world). Figures 7.5 and 7.6 show these two examples.

After pressing the jump control, a player can also build using the build control. This places the note below where the player was before they pressed jump. Fig. 7.9 shows this happening. In the second bar, the F in the treble clef is the player, the E is a permanent note the player has built. The player will remain at F until they jump or move. A player can also "double" jump which means they press the jump control twice in quick succession. This causes them to jump two pitches up. The result is shown in Fig. 7.10. When this is combined with pressing the build control after the double jump, the result is shown in Fig. 7.11.

A player can jump. Pressing the jump control will cause the player's note to move up one key pitch for one cycle of the world and then return back to the previous key pitch. This is shown in Fig. 7.7. A player can also build or place "blocks"—i.e. notes. Pressing the appropriate control will cause a permanent note to be placed at the left or right of the player, as seen in Fig. 7.8. These blocks can be a different timbre to the player's timbre.

If the player jumps and then presses the place block control straight afterwards a block will be created on the line they were just on. This will cause the player to land on that block and not fall down to their original position, as seen in Fig. 7.9. Musically jump can create intervals of a second or minor section. Double jump allows the creation of thirds. Double jump involves the user pressing the jump control twice in rapid succession, and the result is shown in Fig. 7.10. If they press the build control straight afterwards a block will be placed, but a note higher than usual. Fig. 7.11 shows a player using jump and double jump, and placing blocks both times.

Fig. 7.6 Player moves to the left in the world

Fig. 7.7 Player jumps up one note, for one cycle of the world

Fig. 7.8 Player builds wall to right of themselves

Fig. 7.9 Player jumps up one note, and builds a note under themselves

Fig. 7.10 Player double jumps up two notes

The next element to be described in Musicraft is "mobs". Mobs are the monsters of Minecraft. In Musicraft a mob is a note or notes, in a mob timbre. Fig. 7.12 shows a one note mob appearing in the world, and then moving towards the player over two cycles. The player does not move as the mob approaches. In the fourth world cycle the mob destroys the player. Given that timbres cannot be "seen", letters have been used above the bars to clarify. Note how the player in Fig. 7.12 is faster moving than the player (not the case with all mobs). The player can only move one beat a cycle, but the mob can move two. So how can the player escape the mob? Fig. 7.13 shows how. The player jumps and builds a wall, then moves behind the wall by wrapping around the world. Then they build another wall. They are safe from the mob.

A player starts with only a certain number of such "wall" notes available. A little like a Minecraft blocks inventory in survival mode. The player can mine notes beneath themselves to add to the inventory. So the player would need to be part of a chord, harmony or genreto do that.

Looking at this basic functionality of Musicraft it is clear to see how the creative mode allows a user to create their own musical loops. If the mobs cease to be dangerous to the player, then the player can simply build and dig notes wherever they wish. The mobs become non-deterministic features of the tune. Though the user can bound their behaviour by building walls. In fact in Musicraft creative mode, as in Minecraft, the user can spawn mobs. This allows the user to embed dynamic sonic objects of different timbres into their loop.

One advantage that games which simulate a physical environment, or which simulate dance or musical playing, and which Musicraft doesn't have, is that there

Fig. 7.11 Player double jumps up two notes and builds wall

Fig. 7.12 Player pursued and killed by mob, p = player, m = mob

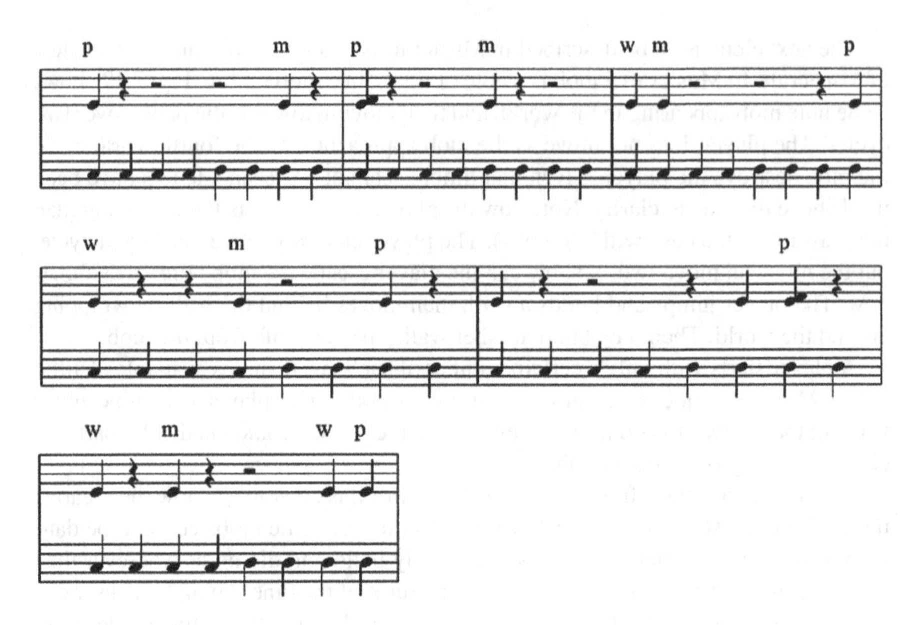

Fig. 7.13 Player protected from mob by building wall, p = player, m = mob, w = wall

is a sense of persistence. This comes from the modelled world seeming natural. The idea of a world existing in a musical loop is less natural. Musicraft is so abstracted compared to these, that it can be expressed as a loop-based music/rhythm game instead of a mob and wall building game. In this view the player controls a note cursor that can be used to place notes. The player can move their note to earlier and later in the loop. The basic game is: sometimes the player hears notes that are coming earlier and earlier each time the loop happens. In this case the player needs to place a note at the same pitch as this note or be destroyed.

There are however potential advantages to thinking in terms of "mobs" and "walls". One is that it provides an opportunity to teach the basics of common music notation. But more importantly it provides a creative metaphor for generating new game features, as will now be demonstrated.

Extending Musicraft

A first obvious element of Minecraft that can be mapped to Musicraft is different block types. Some Minecraft block types are shown in the right hand part of Fig. 7.14. In Musicraft, rather than differing by colour and texture, block types differ by note length and note timbre. Examples of note lengths are crotchets, quavers, semi-quavers, and minims. Timbres include piano, bass drum, violin, snare drum, electric organ, vibraphone, any synthesized sound, and even singing. In fact, one only needs to think of the vast number of sounds available in classical and contemporary music across the world, to the see that the opportunities for sound blocks are vast.

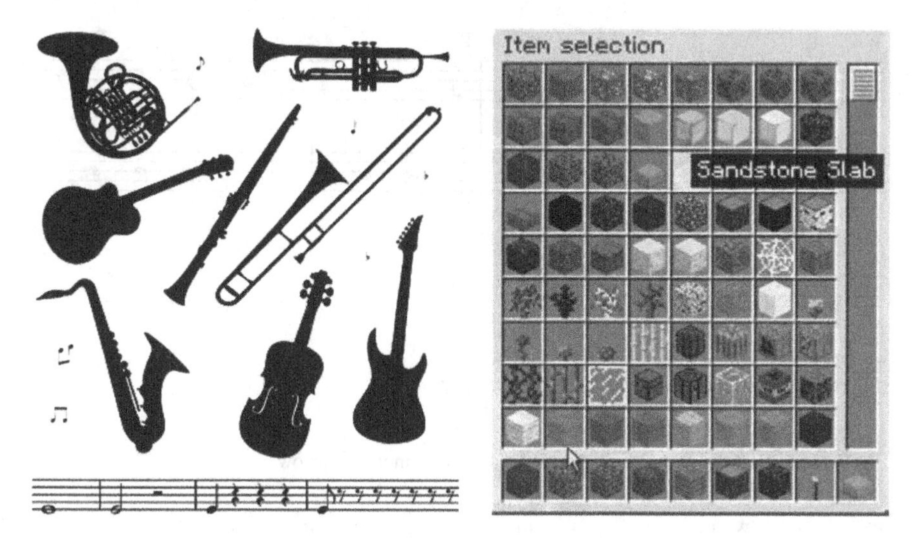

Fig. 7.14 Some Minecraft block types on the right with the sort of items that can be used to build Musicraft blocks on the left

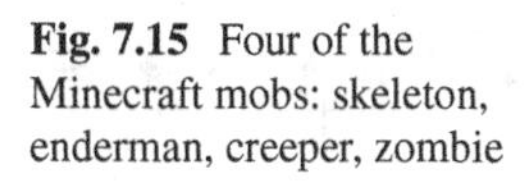

Fig. 7.15 Four of the Minecraft mobs: skeleton, enderman, creeper, zombie

Minecraft also has a variety of mobs which look and behave differently and have varying abilities. Fig. 7.15 shows four of these mobs: the skeleton, the enderman, the creeper, and the zombie. The zombie is the closest to the Musicraft mob described earlier. It simply moves towards the player and kills the player by "touching" them. In Minecraft, the skeleton fires a bow and arrow—so has a ranged attack. The enderman can pick up blocks and can teleport between locations almost instantly. If a creeper gets too close to a player it explodes a few second later, destroying an number of blocks around it.

The skeleton in Musicraft is shown in Fig. 7.16. The skeleton has its own timbre and sends a deadly musical note along a random line near itself. It can be seen in Fig. 7.16 that the skeleton to the far end of the loop is firing an "arrow" made of a quarter note. Luckily for the player, the note misses them. Had the skeleton fired it along middle E, the player would have died unless they avoided it or built a wall. An Underman can be seen in action in Fig. 7.17. It is teleporting around the loop, and in the last bar it removes a block from the player's wall.

Figure 7.18 demonstrates the behavior of a Crawler. In this case the player allows it to get too close, and this triggers the explosive countdown of the Crawler. Then a

Fig. 7.16 Skeleton mob in Musicraft, p = player, m = mob, a = arrow

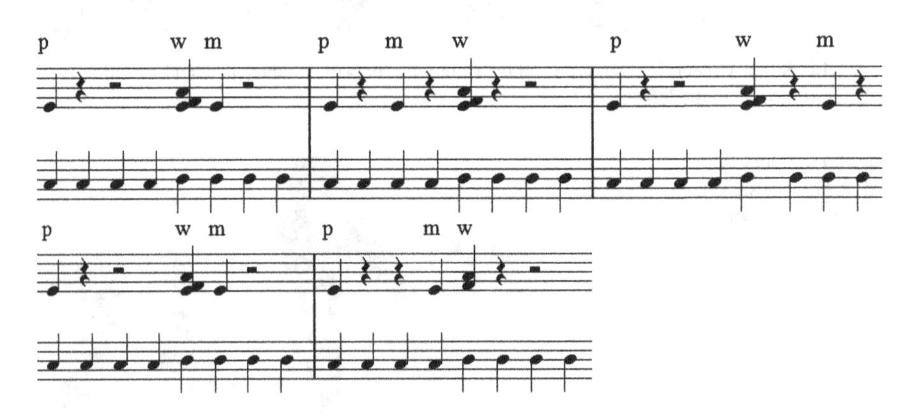

Fig. 7.17 Underman in Musicraft, teleporting and removing blocks

little later the crawler explodes, destroying the nearby wall. The player will hear the Crawler has been activated by a change in timbre.

Minecraft also has a concept of Biomes. These are environments such as plains, desert, and savannah. Fig. 7.19 shows a few Minecraft Biomes. A related concept in Musicraft is genre. Figures 7.20 and 7.21 give two examples of a player building a wall, one in a dance music genre and one in a classical music genre. In both cases the user is in the lowest line, and the other lines are accompaniment.

There are many other extensions which can be envisioned for Musicraft that have stronger or weaker connections with Minecraft. For example multi-player worlds, in which two people control a note each; objects and mobs the move in uniquely musical ways—such as arpeggios, or chords; armor (moveable walls) and weapons (note-structures that damage mobs), ceilings (notes that repeat across time), the ability for players to move between parts of a genre and play there—i.e. into the drum section or the bassline, and finally active note structures which are comparable to "redstone" blocks in Minecraft. Redstone in Minecraft allows the building of

Fig. 7.18 Crawler in Musicraft, exploding and removing a wall

Fig. 7.19 Three Example Minecraft Biomes

Fig. 7.20 Musicraft genre of dance music, showing a player building a wall using double jumps, to avoid an approaching mob

Fig. 7.21 Musicraft genre of classical music, showing a player building a one note wall, to avoid an approaching mob

transmitting and conditionally active structures. They can be used to build computers or factories. However, before implementing such extensions, some core issues need to discussed with moving audiogames and Minecraft-type ideas into a world of common music notation.

Discussion

The key issue with Musicraft is the same as the issue with many audio games: degrees of freedom and persistence. There is a limit to how many different sounds the ear can work with in an auditory scene, and—by its very nature—sound is usually temporary. It fades away unless the sound is continuously played. Visual games can take full advantage of the high resolution visual modality, together with its natural persistence over time. Musicraft is particularly vulnerable to this limitation as it represents the space as a loop. Thus the when the second part of the loop is playing the first half is no longer playing. So to increase the size of a Musicraft world horizontally requires the player to increase their memory model of the space. If rather than 8 beats, it is increased to 16 beats, then the player will need to increase the size of their model. Also the larger the size of the space, the longer it will be before the player can either do something, or hear the results of their actions. So there needs to be a balance between the space size being too large, unresponsive and unwieldly, with the space being too short and low resolution. Of course this can be somewhat offset through the use of tempo. Increasing the tempo of a Biome means it refreshes more quickly. But it also means the player may have to listen more carefully and react more quickly to the space.

Other ways to deal with the resolution issue is allowing play to move between instruments in a Biome, thus increasing freedom in the vertical dimension. Yet more ways include the use of structures whose length is the width of the world. For example an 8 beat note in an eight beat world. This structure is in essence persistent. In fact one can envisage rotating the whole of a Musicraft world 90 degrees and making object locations pitch-based rather than time based. Then a user can hear the whole content of the space constantly. This has the issue that pitch and timbre are likely to clash more in a single time instant, than across multiple beats.

These issues do not remove from the main and the secondary contributions envisaged by the concept of Musicraft to audiogame design. The secondary contribution is the way that whole game forces the user to think in terms of common music notation. For a sighted user this can be a way of learning the basic structures of common music notation without even realizing it. The primary contribution is the idea of bringing a creative mode to audiogames. Through the use of wall building, mob spawning, and many easy-to-imagine tools in Musicraft, it can be seen how once the user has learned the skills in competitive survival gameplay, they can easily transfer these to real-time composition.

Summary

This chapter has discussed the concept of audiogames—where the entire focus of the game is its sound—and introduced a new game concept for audiogames using an example. The concept is of the game being entirely made up of its musical soundtrack—the game is the soundtrack, and of having a creative mode. In particular, this game called Musicraft, is inspired by Minecraft to have both a survival and a creative mode. This is a natural extension of a game that is essentially a piece of music, to allow the player to take the skills they learned in playing the game, to actually creating new pieces of music with those same skills.

References

Allain, K., Dado, B., Van Gelderen, M., Hokke, O., Oliveira, M., Bidarra, R., Gaubitch, N.D., Hendriks, R.C., Kybartas, B.: An audio game for training navigation skills of blind children. In: Serafin, S., Avanzini, F., Geronazzo, M., Erkut, C., de Goetzen, A., Nordahl, R. (eds.) 2015 IEEE 2nd VR Workshop on Sonic Interactions for Virtual Environments (SIVE), pp. 1–4. IEEE, Piscataway, NJ (2015)

Atkinson, M.T., Gucukoglu, S., Machin, C.H., Lawrence, A.E.: Making the mainstream accessible: redefining the game. In: Proceedings of the 2006 ACM SIGGRAPH symposium on Videogames, Boston, MA, 30–31 July, pp. 21–28. ACM, New York (2006)

Beksa, J., Fizek, S., Carter, P.: Audio games: investigation of the potential through prototype development. In: A Multimodal End-2-End Approach to Accessible Computing. Springer, London (2015)

Beksa J., Garkavenko A., Fizek S., Vodanovich S., Carter P.: Adapting videogame interfaces for the visually impaired: a case study of audio game hub. In: 25th International Conference on Information Systems Development, Poland (2016)

Brammer, J.: Audio screen: Unsighted game mechanics for mobile devices. Doctoral dissertation, University of Washington (2016)

Brieger, S.: Sound hunter: Developing a navigational hrtf-based audio game for people with visual impairments. In: Proceedings of 2013 the Sound and Music Computer Conference (2013)

Consalvo, M.: Using your friends: social mechanics in social games. In: Proceedings of the 6th International Conference on Foundations of Digital Games. ACM (2011)

Csapó, Á., Wersényi, G., Nagy, H., Stockman, T.: A survey of assistive technologies and applications for blind users on mobile platforms: a review and foundation for research. J. Multimodal User Interfaces. **9**(4), 275–286 (2015)

Denham, J., McComas, H.: Not just playing around: a review of accessible windows-based games. Access World Mag **5**(5), American Foundation for the Blind (2004)

Dwyer, P., VanLund, P.: The last crusade. https://www.audiogames.net/page/news_55 (2008). Accessed Dec 2017

Fizek, S., Woletz, J.D., Beksa, J.: Playing with sound and gesture in digital audio games. In: Proceedings of 2015 Man and Computer Conference (2015)

Frank, M., Zotter, F., Sontacchi, A.: Producing 3D audio in ambisonics. In: Proceedings of 57th Audio Engineering Society Conference, Audio Engineering Society (2015)

Glinert, E., Wyse, L.: AudiOdyssey: an accessible video game for both sighted and non-sighted gamers. In: Proceedings of the 2007 conference on Future Play. ACM (2007)

Kramer, G.: Auditory Display: Sonification, Audification, and Auditory Interfaces. Perseus Publishing, New York (1993)

Mangiron, C., Zhang, X.: Game accessibility for the blind: current overview and the potential application of audio description as the way forward. In: Researching Audio Description. Springer, London (2016)

McCrindle, R.J., Symons, D.: Audio space invaders. In: 3rd International Conference on Disability, Virtual Reality and Associated Technologies, Alghero, Italy, 23–25 September (2000)

Oren, M.A.: Speed sonic across the span: building a platform audio game. In: CHI'07 Extended Abstracts on Human Factors in Computing Systems. ACM, New York, NY (2007)

Östblad, P.A., Engström, H., Brusk, J., Backlund, P., Wilhelmsson, U.: Inclusive game design: audio interface in a graphical adventure game. In: Proceedings of the 9th Audio Mostly: A Conference on Interaction with Sound. ACM (2014)

Parker, J.R., Heerema, J.: Audio interaction in computer mediated games. Int. J. Comp. Games Technol. **2008**, 178923 (2008)

Rebok, G.W., Ball, K., Guey, L.T., Jones, R.N., Kim, H.Y., King, J.W., Marsiske, M., Morris, J.N., Tennstedt, S.L., Unverzagt, F.W., Willis, S.L.: Ten-year effects of the advanced cognitive training for independent and vital elderly cognitive training trial on cognition and everyday functioning in older adults. J. Am. Geriatr. Soc. **62**(1), 16–24 (2014)

Rovithis, E.: A classification of audio-based games in terms of sonic gameplay and the introduction of the audio-role-playing-game: Kronos. In: Proceedings of the 7th Audio Mostly Conference: A Conference on Interaction with Sound. ACM (2012)

Rovithis, E., Floros, A., Mniestris, A., Grigoriou, N.: Audio games as educational tools: design principles and examples. In: Proceedings of 2014 Games Media Entertainment. IEEE (2014)

Rumsey, F.: Spatial Audio. CRC Press, Boca Raton (2012)

Sánchez, J., Baloian, N., Hassler, T., Hoppe, U.: Audiobattleship: blind learners collaboration through sound. In: CHI'03 Extended Abstracts on Human Factors in Computing Systems, 05–10 April, pp. 798–799. ACM, New York (2003)

Short, D.: Teaching scientific concepts using a virtual world—Minecraft. Teach. Sci. **58**(3), 55 (2012)

Spiel, K., Bertel, S., Heron, M.: Navigation and immersion of blind players in text-based games. Comp. Games J. **3**(2), 130–152 (2014)

Velleman, E., van Tol, R., Huiberts, S., Verwey, H.: 3D shooting games, multimodal games, sound games and more working examples of the future of games for the blind. In: Proceedings of 2004 International Conference on Computers for Handicapped Persons. Springer, Berlin Heidelberg (2004)

Westin, T.: Game accessibility case study: terraformers–a real-time 3D graphic game. In: Proceedings of the 5th International Conference on Disability, Virtual Reality and Associated Technologies, ICDVRAT 2004 (2004)

Westin, T., Bierre, K., Gramenos, D., Hinn, M.: Advances in game accessibility from 2005 to 2010. In: Universal Access in Human-Computer Interaction. Springer, Heidelberg (2011)

Yuan, B., Folmer, E.: Blind hero: enabling guitar hero for the visually impaired. In: Proceedings of the 10th International ACM SIGACCESS Conference on Computers and Accessibility. ACM, New York, NY (2008)

Yuan, B., Folmer, E., Harris, F.C.: Game accessibility: a survey. Univ. Access Inf. Soc. **10**(1), 81–100 (2011)

Chapter 8
Motion Controllers, Sound, and Music in Video Games: State of the Art and Research Perspectives

Federico Visi and Frithjof Faasch

Current State of the Art

Motion-Based Video-Gaming Controllers

This section describes recent (currently available or in advanced development stage) motion-based controllers and their use in gaming. The devices are subdivided in two main categories: handheld controllers and camera-like controllers. The former includes devices that are held in the player's hands, while the latter consists in devices that capture the player's movement without requiring direct manipulation. These two typologies employ various motion-sensing technologies and also differ in terms of the interactions they can afford, which have implications for the design of motion-based gameplay mechanics. The section does not aim at providing an exhaustive list of all the available controllers. Rather, its purpose is to look at the motion-sensing technologies that are more frequently employed in popular gaming platforms and therefore characterise the design of the motion-based interactions.

Handheld Motion Controllers

Handheld motion-sensing game controllers have become more popular since Nintendo introduced the Wii Remote (Fig. 8.1a) in 2006. This kind of controllers often have built-in inertial sensors, such as accelerometers and gyroscopes. Accelerometers allow the tracking of translational acceleration of whatever they are attached to, and are often found in 3D configurations (i.e. able to sense acceleration along three orthogonal axes). Gyroscopes, on the other hand, allow to measure

F. Visi (✉) • F. Faasch
Universität Hamburg, Hamburg, Germany
e-mail: mail@federicovisi.com; frithjof.faasch@web.de

D. Williams, N. Lee (eds.), *Emotion in Video Game Soundtracking*, International Series on Computer Entertainment and Media Technology,
https://doi.org/10.1007/978-3-319-72272-6_8

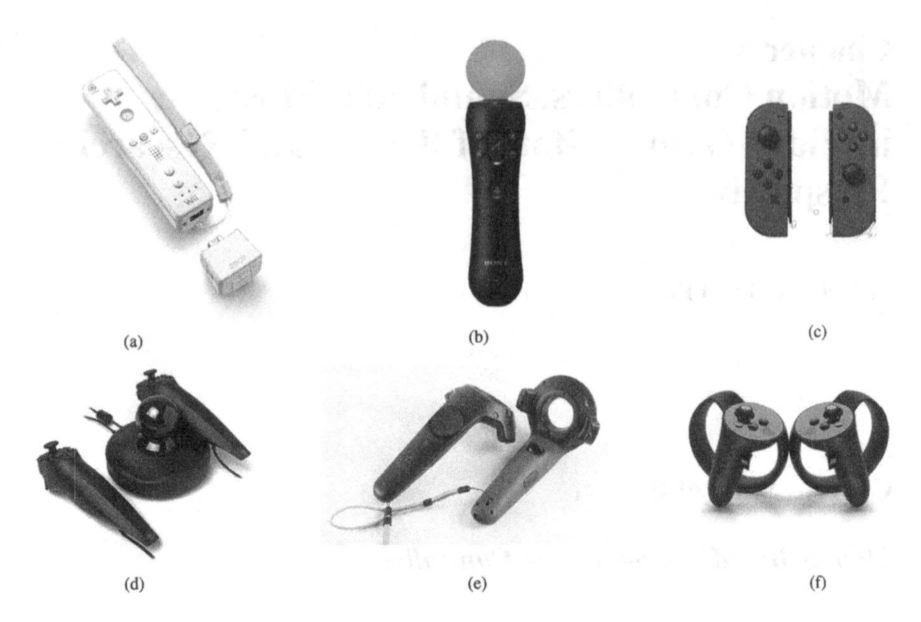

Fig. 8.1 Handheld motion controllers: (a) Wii Remote with Wii MotionPlus expansion; (b) PlayStation Move; (c) Joy-Con; (d) Razer Hydra; (e) Vive Controller; (f) Oculus Touch

rotational velocity and are also frequently found in 3D configurations. These two sensors are used in tandem in inertial measurement units (IMUs), which are used extensively in aviation, robotics, Human–Computer Interaction (HCI), and for health applications such as gait analysis (Trojaniello et al. 2014; Visi et al. 2017c). In recent years, their increasing affordability and small size have made them a common feature of mobile and wearable devices and other consumer electronics. Due to drift error, inertial sensors do not currently allow accurate position tracking in real time. To overcome this limitation, some handheld motion controllers also include other motion-sensing technologies, such as optical or magnetic sensors. Principles for using accelerometers in game controllers, including operating mechanisms and specific parameters that directly affect gaming performance, are discussed by Champy (2007).

When it was introduced in 2006, the only inertial sensor in the Wii Remote was a 3D accelerometer. In 2009, Nintendo released an expansion device for the controller named Wii MotionPlus (Fig. 8.1a), which included gyroscopes and could be attached to the original controller to allow more accurate motion tracking. The following year, Nintendo started producing a new version of the Wii Remote with built-in gyroscopes that provided the same motion-sensing capabilities of the Wii MotionPlus without the need of an expansion module. In addition to the inertial sensors, the Wii Remote has an optical sensor on the top end. This is used together with the Sensor Bar, a device with two groups of infrared LEDs to be placed above or below the screen. Since the distance between the two group of LEDs is known, it is possible to estimate the direction the Wii Remote is pointing to through triangulation when the Sensor Bar LED's are captured by the Wii Remote optical sensor. In addition to the Wii MotionPlus, Nintendo has introduced various other accessories and expansions for the Wii Remote. The Nunchuk connects to the Wii Remote through a cable and it is supposed to be held with the other hand by the player. It contains a 3D accelerometer similar to the one found in the Wii Remote, but no gyroscopes.

There are other accessories for the Wii Remote such as the Wii Wheel (a wheel with a slot for a Wii Remote), the Wii Zapper (a gun-shaped shell that can house a Wii Remote and a Nunchuk), and various others including a guitar-shaped device with a slot for a Wii Remote. These do not provide additional motion sensing capabilities, but are still of interest since they are designed to provide different affordances and ergonomics, thus suggesting other uses and movements to the player.

The PlayStation Move (Fig. 8.1b) released by Sony in 2010 is similar to the Wii Remote as it also includes accelerometers and gyroscopes for inertial motion sensing. However, optical sensing in the PlayStation Move works differently since there is no optical sensor on the controller itself. At the top of the controller there is instead an orb that emits light of different colours using RGB LEDs. A camera placed near the screen tracks the position of the orb, and (since the size of the orb is known) the system is capable of estimating the distance between the controller and the camera by tracking the size of the orb in the camera's image plane. Before releasing the PlayStation Move, Sony updated its DualShock gamepad by adding inertial sensors to it. This is a feature found also in other more recent gamepads such, as the Steam Controller developed by Valve, although these devices will not be covered in this section since they are primarily designed as gamepads and not as motion controllers.

After the commercial success of the Wii Remote, Nintendo included motion sensors in their following two consoles. Both the Nintendo 3DS handheld console and the gamepad of the WiiU home console include a 3-axis accelerometer and a 3-axis gyroscope. However, compared to their predecessor, the games produced for these consoles employed motion control only to a limited extent.

More recently, Nintendo has introduced the Joy-Con (Fig.8.1c), the main controller of the Nintendo Switch game console. Released in 2017, the Joy-Con improves and extends many of the concepts and technologies that characterised the Wii Remote and also allows for a more traditional gamepad-like configuration, achieved by attaching the controllers to the game console itself or to an accessory grip. Joy-Con come in pairs, with a controller designed to be used mainly with the right hand and a second one for the left hand ('Joy-Con R' and 'Joy-Con L' respectively). This allows the Joy-Con to be used as a two-handed motion controller, similarly to a Wii Remote paired to a Nunchuk. More compact and light than a Wii Remote, each Joy-Con contains a 6-axis inertial measurement unit (IMU) comprising both accelerometers and gyroscopes (STMicroelectronics 2017). The direction the Joy-Con is pointing to can be tracked without the need of an additional external device (as it was the case with the Wii Remote and the Sensor Bar). This is possible after a quick calibration procedure[1] requiring the Joy-Con to rest on a flat surface (Nintendo 2017; Kuchera 2017). The pointer can be re-centred at any time using a specific button, suggesting that it is not necessary to point the controller at the screen to make direction tracking work (Kuchera 2017). Additionally and unique to the Joy-Con R, an infrared depth sensor placed on one end of the controller allows to detect the distance and shapes of nearby objects, and can also be used for hand gesture recognition purposes (Takahashi 2017).

The Razer Hydra (Fig.8.1d) controller developed by Sixense Entertainment in partnership with Razer and released in 2011 presents some unique features in terms of motion tracking technology. With a motion sensitive unit for each hand equipped with buttons and direction sticks, the Razer Hydra is functionally similar to other handheld motion controllers. However, it differers from the Wii Remote and the PlayStation Move as it does not use optical means for position tracking. Instead, it employs low-power magnetic fields, allowing 6 degrees of freedom motion tracking (position and orientation) of both controllers without the need of calibration procedures. The base station that senses the movements of the controllers does not require unobstructed line-of-sight, as it would be the case with optical sensors. According to its developers (Sixense Entertainment 2011), the technology allows low-latency tracking with precision of 1 mm and 1°. However, magnetic field tracking might suffer from the interference of other magnetic fields emitted by nearby devices. A downside of the Razer Hydra in comparison to other handheld motion controllers is the presence of wires. At the time of writing though, Sixense has made available to developers a wireless system named STEM. It employs magnetic field motion tracking similarly to the Razer Hydra and it is mainly targeted at virtual reality (VR) applications.

Recent handheld controllers dedicated to virtual reality such as the Vive Controllers (Fig. 8.1e) and the Oculus Touch (Fig. 8.1f) are functionally similar to other motion controllers described here as they are also equipped with buttons and sensors. However, since virtual reality applications often require precise and consistent positional tracking, these controllers rely on multiple external infrared optical sensors, which are used also to track the position of VR head-mounted displays (HMDs).

A new device currently being developed by Valve might introduce some novelty in the panorama of handheld motion controllers. Known as the Knuckles, these devices add finger tracking capabilities to inertial and positional tracking common to other controllers. The Knuckles are strapped on the hands, allowing the user to move all the fingers without risking of dropping the devices (Yang 2017b). Several capacitive sensor placed on the handgrip track the movement of individual fingers (Yang 2017a), potentially allowing detailed smaller-scale interactions. At the time of writing, Valve is distributing the controllers to a limited number of selected developers.

[1]This calibration procedure suggests that direction tracking is achieved by using drift-corrected orientation data obtained from the built-in IMU. A similar calibration procedure was previously used also in games employing the Wii MotionPlus expansion for the Wii Remote.

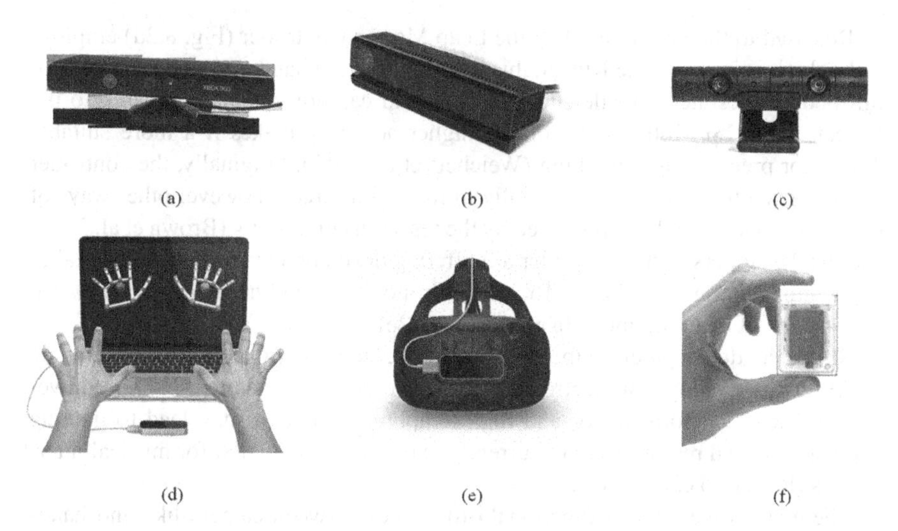

Fig. 8.2 External Camera-like Motion Controllers: (**a**) Kinect (version 1); (**b**) Kinect (version 2); (**c**) PlayStation Camera (2016 version); (**d**) A Leap Motion Controller used as a tabletop device with a laptop; (**e**) A Leap Motion Controller mounted on a head-mounted display; (**f**) Soli Development Kit

External Camera-Like Motion Controllers

This section presents an overview of motion controllers that do not require the user to directly hold the device in order to work. These usually consist in devices that work similarly to a camera: they are placed nearby and pointed to the players in order to track their movements. However, these devices often employ other technologies in addition to optical sensing. The Kinect developed by Microsoft is probably one of the most popular camera-like motion controller for gaming applications. The first version introduced in 2010 (Fig. 8.2a) includes an RGB camera and an active depth sensor (thus resulting in what is also known as an RGB-D sensor). The latter works by projecting an invisible pattern through an infrared projector, which is then captured by a monochrome infrared sensor. Depth data is reconstructed by analysing how the projected pattern is reflected by surrounding objects and bodies. By combining the information obtained from its sensors, the Kinect is able to track the position of specific points on the player's full body, called joints. The second iteration of the controller, Kinect 2.0 (Fig. 8.2b), was released in 2013 and brought many improvements over the first Kinect, particularly in terms of tracking accuracy, resolution, and field of view (Wasenmuller and Stricker 2017).

After releasing the simpler PlayStation Eye (a device similar to a webcam with limited motion tracking capabilities based on computer vision and gesture recognition algorithms) in 2007, Sony released the PlayStation Camera, which—differently from the Kinect—uses stereoscopic vision to track depth. The first version released in 2013 was followed by an update in 2016 (Fig. 8.2c), which included slight improvements (Halston 2016).

Released to the public in 2013, the Leap Motion Controller (Fig. 8.2d) employs technologies similar to the Kinect, this time for tracking hand movements instead of full bodies. Despite a smaller field of view and capture volume compared to the Kinect, the Leap Motion Controller's higher accuracy makes it a more suitable device for precise finger tracking (Weichert et al. 2013). Originally, the controller was designed to be placed horizontally in front of a screen, however other ways of using the device have been proposed by the research community (Brown et al. 2016) and the developers of the controller are offering development kits for virtual reality applications that use the Leap Motion Controller as a hand tracking device placed on the back of head-mounted displays (Fig. 8.2e) (Holz 2014).

Still under development at the time of writing, Google's SOLI (Fig. 8.2f) (Google ATAP 2017) employs an electromagnetic wave radar to capture very fine movements. The small form factor and high temporal resolution may lead to gaming applications, and researchers are currently exploring its potential for musical interactions (Bernardo et al. 2017).

There are some evident functional differences between camera-like and handheld motion controllers, the most evident of which is that the former does not require the player to manipulate a device in order work. However, differences go beyond that, as different systems represent movement through motion data in different ways. This has some crucial implications on how meaningful movement interactions are designed, as it will be described in section "Recent Research in Expressive Movement Analysis: Motion Descriptors and Mapping".

Examples of Relationships Between Motion Control, Sound, Music, and Gameplay Mechanics

Following the review of the hardware devices available, we present and discuss some gameplay mechanics involving motion, sound, and music, with particular attention to the representation of musical instruments in video games and the design strategies adopted for implementing musical interactions.

Rhythm games or rhythm-action games are 'video games in which the player must respond in some way to the rhythm or melody being presented, either through repeating the same melody or rhythm by pressing buttons (with hands or feet), or kinetically responding in some other way to the rhythm, often using specially designed controllers' (Collins 2008, p. 74). The first examples of this game genre date back to the late 1970s (Collins 2008). The category gained renewed popularity between the late 1990s and the 2000s, with hits such as Dance Dance Revolution and Guitar Hero. Both games involved the use of specially designed controllers: the former employs a dance pad with panels the player is supposed to step on following the dance patterns presented on the screen, whereas the latter a guitar-shaped controller (more about this type of controllers in section "In-Game Musical Instruments"). More recently, music-themed rhythm games started taking advantage of motion controllers. The dance game Just Dance introduced by Ubisoft in 2009 eschewed the use of dedicated devices in favour of the Wii Remote. In the game, players follow the dance sequences performed on screen by an animated dancer while holding a Wii Remote. They receive points for how accurately they mimic the dance moves of the on-screen characters. The first episode received generally unfavourable reviews from the critics (CBS Interactive 2017b), as its gameplay was considered excessively basic and motion detection was deemed imprecise. However, the game was a commercial success and users praised its simplicity and multiplayer mode that allowed four people to play at the same time (CBS Interactive 2017b). The third instalment of the series released in 2011 introduced the support for Kinect and PlayStation Move, while later episodes gradually improved motion tracking and also included support for Nintendo Switch and smartphones equipped with motion sensors. Following the success of Just Dance, in 2010 Harmonix (the development company behind other well-known music-themed rhythm games such as Guitar Hero and Rock Band) released Dance Central. Available exclusively for Microsoft's Xbox 360, Just Dance takes advantage of the full-body motion capabilities of the Kinect. According to critics (CBS Interactive 2017a), this resulted in a rewarding gaming experience, with more accurate motion tracking and variable difficulty, allowing to go beyond casual playing and increasing the longevity of the game. Compared to Just Dance, Dance Central offers the player more possibilities to improve their skills. Full-body motion tracking allows the game to show which dance moves can be improved in detail and a training mode allows to break dance sequences down to individual moves and practice at regular speed or in slow motion.

Close relationships between music, gameplay mechanics, and motion control are not found exclusively in rhythm games. For example, in some stages of Nintendo's Super Mario Galaxy (2007), the player controls Mario while the character is balancing on a sphere. By tilting the Wii Remote, the player makes Mario walk on the sphere. The direction and angle of the tilt correspond to the direction and speed of Mario's walk. The instability of this balancing act is underlined by the music, which changes dynamically according to the sphere movements. The speed of the melody follows the velocity of the sphere, and so does the timbre of the whole tune. This results in the music sounding as if it was being played back by a tape recorder at a speed that constantly changes with the dynamics of the movements of the character on screen. This contributes, albeit subtly, to the sense of unsteadiness stemming from Mario's movements, and helps the player to anticipate and follow the dynamics of the interaction.

Multimodal relationships and links between music and gameplay are more pronounced in Child of Eden, developed by Q Entertainment and published by Ubisoft in 2011. It is compatible with Kinect and PlayStation Move, but can also be played using traditional controllers. In the game, the player aims at various targets using one or more on-screen pointers (the 'Reticles'). When using motion controllers, the player controls the Reticles with the movement of their hands. Different reticles have different gameplay mechanics: the blue one has a lock-on function and the player can release a laser shot by quickly flicking their hands when locked on a target, while the other reticle is characterised by a rapid fire function that is more effective with certain targets. On-screen visuals are synchronised to the music and hitting a target often results in sounds and musical effects that match the key or tempo of the backing soundtrack. If using a Kinect, the player can swap the control of the two Reticles between left and right by clapping their hands. Since all action on screen is synchronised to a beat-heavy music, this and other gestures involved in controlling the Reticles to hit targets contribute to build up a very rhythmical, fast-paced gameplay tightly coupled with sounds, music, and visual effects.

Another example of gameplay mechanics with strong interplay between music and motion can be found in the game Johann Sebastian Joust,[2] part of the game collection Sportsfriends released by the independent developer Die Gute Fabrik in 2014. Johann Sebastian Joust offers some rather distinctive gaming dynamics, which perhaps even stretch beyond the conventional definition of video game. It is a local (i.e. not networked) multiplayer game with no on-screen visuals specifically designed for handheld motion controllers. Its designers describe it as 'a duelling game, controlled by the tempo of the music'.[3]

The main gameplay mechanics are effectively summarised by the developers on the game's web page:

> The goal is to be the last player remaining. When the music – J.S. Bachs's Brandenburg Concertos – plays in slow-motion, the controllers are very sensitive to movement. When the music speeds up, the threshold becomes less strict, giving the player a small window to dash at their opponents. If your controller is ever moved beyond the allowable threshold, you're out! Try to jostle your opponents' controllers why protecting your own. (Die Gute Fabrik 2014)

Interestingly, the developers used, perhaps unwittingly, the term 'slow-motion' instead of slow tempo, possibly anticipating the effect that the music will have on the movements of the players. In the game, the music is front and centre, as it regulates the pace of each match by letting the players know which rules are currently in play, that is, how tolerant to sudden movements their controller is. As the tempo of the music increases, so does the tolerance of the motion controllers, which allows players to move more rapidly and possibly attempt to reach and set off other players' controllers. If the controller of one of the player is shaken beyond the acceleration threshold, he or she is eliminated from the game, living the remaining players 'jousting' for the final victory. Albeit very essential, the game provides an effective example of successful gameplay mechanics involving no screen, only music, sound, and motion controllers.

In-Game Musical Instruments

An *in-game musical instrument* is a representation of a musical instrument *within* the game world that the character controlled by the player can interact with. This has conceptual and practical differences from music rhythm games involving

[2] http://www.jsjoust.com.

[3] From the video trailer of the game (at 00:13) on Die Gute Fabrik Vimeo channel: https://vimeo.com/31946199.

instrument-like controllers such as Guitar Hero. Such games employ controllers shaped to resemble a musical instrument usually equipped with a set of buttons and other simple input devices placed in a way to induce some instrumental-like gestures. For example, the Guitar Hero controller is shaped like a smaller, stringless guitar with a set of buttons on the fretboard and a two-way directional pad placed in the middle of the guitar body, used to simulate strumming. The main goal of musical rhythm games is to follow visual cues presented on screen and push the corresponding button on the controller in time with the music in order score points. On the other hand, in-game musical instruments are objects represented in the game that the player can use and interact with to produce music with various degrees of freedom.

The action-adventure games of the Legend of Zelda series often include some form of musical interaction. The sophistication of such interactions has varied through the years: from simply pressing a single button to play tones with a recorder in the first episode of the series released in 1986, to a set of five tones playable with the Nintendo 64 controller in Ocarina of Time (1998), to a mini rhythm game to control the direction of the wind with a conductor's baton in The Wind Waker (2003). The latter was further refined with motion controls in the remastered version released for the WiiU in 2013. Moreover, in the episode Spirit Tracks published in 2009, the player can play a pan flute by touching the display of the Nintendo DS while simultaneously blowing in the microphone of the device. Finally, the harp in The Legend of Zelda: Skyward Sword is an effective example of motion-controlled in-game musical instrument. In the game developed and published in 2011 by Nintendo for the Wii console, the player takes the role of the series protagonist, Link. The game makes extensive use of the Wii MotionPlus expansion to the Wii Remote (see section "Handheld Motion Controllers") and offers more sophisticated motion-based gameplay mechanics compared to other Wii games available at the time of release. Similarly to other titles in the Zelda series, music contributes considerably to the gaming experience. Particularly, at some point in the game, Link receives a harp that can be used by the player to play simple melodies and strum chords. This is done by pressing a button on the Wii Remote while waving it in the air left and right. These movements correspond to harp strums, and their speed and timing affects the arpeggiated chords produced by the instrument. At some point in the game, a non-playing character (NPC) teaches the player how to use the harp and then challenges them to play at a specific tempo while he sings along. The phases of this interaction are depicted in Fig. 8.3. First, the player is instructed by the NPC on how to play the harp. After a brief tutorial, the NPC asks the player to strum the harp following the tempo of his swinging braid. If the player does not strum in time, the NPC will act as a "music teacher", telling the player whether they are strumming too fast or too slow. After this "introductory lesson" (top two screenshots in Fig. 8.3), the player has to strum the harp in time with a pulsating circle of light (middle screenshots in Fig. 8.3). If the player stays in time, the NPC will start singing along the harp chords. After a few chords at the right tempo, Link will learn the 'Ballad of the Goddess' (bottom screenshots in Fig. 8.3) and the player will be able to strum the song again later in the game. According to an interview with the game producer

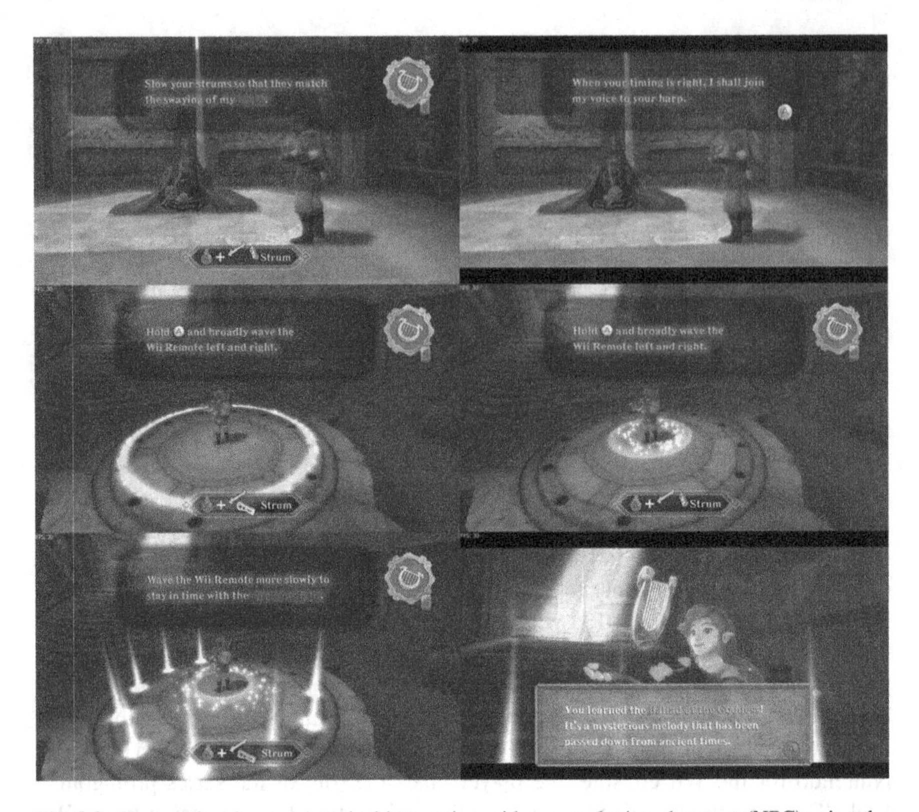

Fig. 8.3 Phases of an in-game musical interaction with a non-playing character (NPC) using the harp in The Legend of Zelda: Skyward Sword

Eiji Aonuma (George 2011), the design team deliberately chose to allow the player to strum the harp whenever they pleased, not only when playing the instrument is required to solve puzzles and advance in the story. In fact, at any point in the game the player can take the harp out and play a few chords in harmony with the background music.

Even though very effective and cleverly designed, the harp in The Legend of Zelda: Skyward Sword gives just a glimpse of the potential and the complexity of musical interactions in video games. In-game musical instruments can be used to advance in the game as well as to play freely whenever the player feels like. Nonplaying characters can serve as music teachers as well as performers to play with. Designing sophisticated and rewarding in-game musical interactions is a complex task that could be made easier by making use of recent research findings in the field of virtual reality. Serafin et al. (2016) discuss the development of musical instruments in virtual environments. They point out that 'virtual reality musical instruments' (VRMIs) differ from 'virtual musical instruments' (VMIs, i.e. software simulations of existing instruments, mostly focused on sonic emulation) as the former also include a 3-D simulated visual component delivered through virtual reality headsets or other immersive visualisation systems. In addition to presenting an overview of the current state of the art and describing case studies and evaluation methods, they propose various principles for the design of VRMIs. They recommend to take into considerations multiple sensory modalities, suggesting that while sound and visuals remain key elements 'touch and motion should also enter into the equation, as well as full body interaction' (Serafin et al. 2016, p. 26). This not only results in a more immersive experience, it also improves the playability of the instrument, allowing the development of more sophisticated musical skills.

Referring to one of Cook's principles for designing computer music controllers stating that 'Copying an instrument is dumb, leveraging expert technique is smart' (Cook 2001, p. 3), Serafin et al. suggest that VRMIs should make use of existing skills and adopting metaphors from existing real-world interactions. On the other hand, attempting to exactly replicate traditional instruments may not bring about interesting results. The limitations of virtual reality would likely make such interactions difficult and, at the same time, designs that try to faithfully replicate traditional instruments may not take advantage of the distinctive possibilities offered by the medium. In fact, the authors encourage designers 'consider the use of both natural and "magical" interactions and instruments' (Serafin et al. 2016, p. 28), that is, not necessarily constrained by real-world physics, technology, and human anatomy. Additional principles put forward by Serafin et al. focus on the representation and presence of the player's body and on the social dimension of music making, stressing the importance of shared experiences and the sense of presence and agency. Many of these guidelines are informed by the output of research communities active in the fields of computer music, human-computer interaction, and new instruments for musical expression (NIMEs, see Jensenius and Lyons 2017, for an overview). Even though these recommendations are specific for the medium of virtual reality, many principles put forward by Serafin et al. can be applied to the design of in-game musical instruments also for non-VR video games, whether the player interacts with the game world in a first-person view or through an avatar. More examples of how interdisciplinary research outcomes can inform the design of musical interactions in video games is further discussed in section "Future Scenarios: Perspectives for Motion-Based Gaming from Current Interdisciplinary Research".

Future Scenarios: Perspectives for Motion-Based Gaming from Current Interdisciplinary Research

This section provides an overview of key topics in motion-based interaction research, such as the extraction of useful motion descriptors from motion data and parameter mapping for expressive interaction. Moreover, we describe recent research projects involving emotion recognition from full-body motion, and the analysis of body movements to study different aspects of player engagement.

Recent Research in Expressive Movement Analysis: Motion Descriptors and Mapping

Human motion research is an increasingly active field, contributing to many areas of study such as behavioural sciences, human–computer interaction, health and rehabilitation, systematic musicology, dance studies, ergonomics, user studies and others. Particularly, numerous interdisciplinary studies centred on embodiment and the experience of music as a multimodal medium have been carried out in the past two decades (Godøy and Leman 2010; Wöllner 2017; Visi 2017). Body movement is also a recurring topic among researchers and designers working computer music interaction and new musical instruments (Jensenius 2014). Analysis of expressive body movement in dance has also been the subject of various studies, some of which employed motion controllers such as the Kinect (Camurri et al. 2004; Alaoui et al. 2012).

In these research fields, the way data obtained using motion-sensing technologies are processed to extract useful information is a key research topic. Various computable motion descriptors have been used for movement analysis and interactive applications. The Eyesweb platform (Camurri et al. 2007) contains a set of tools for motion analysis and is compatible with high-end motion capture system as well as low-cost motion controllers. Among motion descriptors, quantity of motion and contraction index are often used for analysing expressivity in applications involving full-body motion (Glowinski et al. 2011; Visi et al. 2014). The former is used to estimate the total amount of displacement over time, while the latter is an indication of the spatial extent of the body. Extracting meaningful features from motion data is a crucial step in the design of advanced interactions with music and sound. Larboulette and Gibet (2015) recently attempted a thorough review of computable descriptors of human motion, while Piana et al. (2013) described a set of motion features for emotion recognition (see section. "Body Motion, Emotion, Engagement").

The motion controllers described in section "Motion-Based Video-Gaming Controllers" employ a number of technologies that represent movement in different manners. While camera-like systems often describe body movement through positional data of various points of interests (e.g. the joints in Kinect skeletons) the data returned by the inertial sensors in handheld controllers consists in 3-dimensional vectors describing accelerations and rotational velocities, without a global coordinate system. This has several implications for the computation of motion descriptors useful for sound and music interaction. Visi et al. (2017b) present motion descriptors dedicated to inertial sensors, and describe their use in music performance and composition.

Extracting meaningful descriptors from motion data is not the only crucial step in designing successful motion interactions. Mapping them to parameters for controlling sound and/or visuals is another essential aspect of expressive interaction design. Different approaches to mapping have been the subject of numerous studies (Rovan et al. 1997; Van Nort et al. 2014; Hunt et al. 2002). Recently, machine learning has been increasingly employed for creating gesture-to-sound mappings (Caramiaux et al. 2014) and for creating complex mappings from databases of musical gestures (Visi et al. 2017a).

Research on motion descriptors and mapping strategies for musical interaction can inform the design of gestural interactions in gaming. As an example, the collection of game apps developed for the Mogees Play[4] sensor stemmed from the development of a device for musical applications, which itself employed various gesture recognition technologies resulting from academic research (Zamborlin et al. 2014).

Body Motion, Emotion, Engagement

Motion-based gaming has received attention from researchers for different purposes. There are many instances of serious games for healthcare (Wattanasoontorn et al. 2013). Among these, Piana et al. (2016) developed two games designed to help children affected by autism recognise and express emotions through full-body movement. In Guess the Emotion, the player is presented with a short video showing the silhouette or a stick figure of a person expressing a basic emotion (e.g. happiness or anger) through body movement (Fig. 8.4a). Then, the player has to guess which emotion was expressed by the silhouette by picking one from a list of possible answers (Fig. 8.4b). If the answer is correct, the player gains points and the game asks them to express the same emotion through body movement alone (Fig. 8.4c). If the emotion recognised by the system is correct, the player gains additional points, otherwise they will be asked if they want to try to perform that emotion again (Fig. 8.4d). To recognise the emotions expressed by the player, the system uses an RGB-D camera like the Kinect to extract a set of motion descriptors relevant for online emotion recognition. Classification is then performed by a classifier based on Support Vector Machines (SVM) (Burges 1998).

A similar approach is adopted in the second game, Emotional Charades. The game involves two players with different roles: the 'Actor' chooses an emotion and expresses it by moving in front of the sensor. The second player (the 'Observer') will have to guess which emotion was chosen by the Actor by looking at the on-screen silhouette recorded by the sensor. The computer will try to classify the emotion as well, and the guesses of the Observer and the classifier will be shown on screen. The Actor then reveals what the correct answer is, and will gain more points if both the computer and the Observer have guessed right. Similarly, the Observer will gain points if they guessed the emotion expressed by the Actor right. Then the players switch roles.

Bianchi-Berthouze (2013) studies the relationship between body movement and player engagement in motion-based video games. She argues that body movement has a strong influence on the sense of presence of the player and on the overall engagement experienced while playing. By examining the movement patterns

[4] http://www.mogees.co.uk/play.

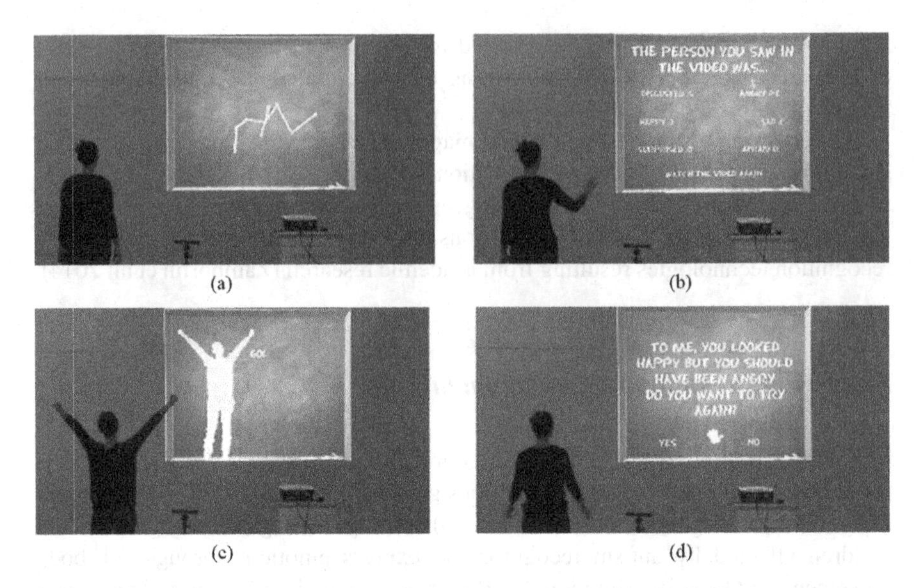

Fig. 8.4 Guess the Emotion, a game by Piana et al. (2016): (**a**) an animated stick figure presents a basic emotion; (**b**) the player tries to guess the emotion; (**c**) if the answer is correct, the player is asked to express the same emotion using body movement; (**d**) the system recognises and evaluates the emotion expressed by the player

players adopt when playing games using both desktop and motion controllers, she proposes a taxonomy of body movements that will be the basis of a hypothetical model of the relationship between player body movement, controller, and player engagement. This taxonomy includes five classes: 'Task-Control Body Movements' (defined by the controller and necessary to control the game), 'Task-Facilitating Body Movements' (performed to facilitate game control but not recognised by the controller), 'Role-Related Body Movements' (typical of the role adopted by the player in the game scenario but not recognised by the controller, e.g. head-banging while playing Guitar Hero), 'Affective Expressions' (gesturing expressing the affective state the player during game play, generally not recognised by the controller), and 'Expressions of Social Behaviour' (movements that facilitate and support interaction between players, not currently recognised by the controller) (Bianchi-Berthouze 2013, pp. 49–51). This movement classes are then related to the four engagement factors previously identified by Lazzaro (2004), resulting in a model useful for a more systematic approach to the design of full-body interactions in video games (Fig. 8.5). To show how motion control can facilitate engagement and result in a broader set of emotions in the player, Bianchi-Berthouze (2013) reports a series of empirical studies focused on a group of participants playing the music-themed rhythm game Guitar Hero with different controllers. Overall, the experimental results showed that, when using a motion-enabled controller, players tend to shift between the engagement types related to the different movement classes described above, showing movements that relate to role-play and enjoyment. On the

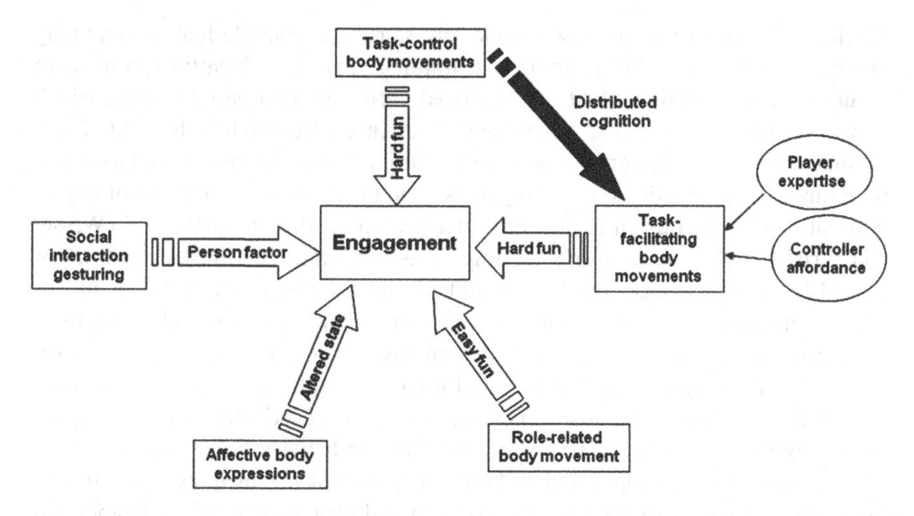

Fig. 8.5 Bianchi-Berthouze's movement-based engagement model for the design of full-body interactions in video games (Bianchi-Berthouze 2013)

other hand, when using a controller that 'do not require and do not afford natural body movements' (Bianchi-Berthouze 2013, p. 60), she observed a general lack of overt movements beyond those necessary to control the game. She then concludes that a more complete and compelling gaming experience involves body movements from all the five classes of the taxonomy, which can also be used as mood and attitude induction mechanisms.

Bianchi-Berthouze's findings (Bianchi-Berthouze 2013) as well as the emotion recognition techniques employed by Piana et al. in their serious games (Piana et al. 2016) constitute useful resources for game designers. Their work on affective states and emotional engagement shows that there is still considerable potential in motion control that has not been fully explored in mainstream video game design.

Discussion and Conclusion

By looking at recent instances of motion control in commercial video games and at the findings of interdisciplinary research involving video games and body movement, one can clearly see that the field is steadily evolving.

Albeit showing increasing sophistication, motion-based interactions in commercial games is still somewhat limited, especially when involving music and sound. Research in neighbouring fields such as computer music and human–computer interaction can lead to useful insight for making musical interaction in games more profound an effective. As seen in the dance games mentioned in section Examples of Relationships Between Motion Control, Sound, Music, and Gameplay

Mechanics", recent motion-based games are starting to include features and play modes that allow for skill-learning and training. This is a departure from more casual gaming modalities that characterised earlier motion-based games, which may allow for more sophisticated gaming experiences. Research in the field of new musical instruments design can provide useful guidelines for obtaining more profound musical interactions in video games. One of the most cited tenets of digital musical instrument design is 'low entry fee with no ceiling to virtuosity' (Wessel and Wright 2001), meaning that getting started with a digital musical instrument should be relatively easy but this should not hinder the development of higher degrees of expressivity. The same principle can be applied to musical interactions in video games, whether these are based on dance or on in-game musical instruments. The learning curve should allow for casual gaming and, at the same time, for the development of virtuosity. This way, progressive skill-learning is possible, increasing the longevity of the game and the depth the musical interactions it affords. Moreover, appropriation and the emergence of personal styles contribute to define the identity of a musical instrument and characterise its use (Zappi and McPherson 2014). Similarly, musical interactions in video games would benefit from the emergence of personal styles, which make the experience more intimate and distinctive for each player.

With the increasing development and availability of virtual reality and augmented reality HMDs, video games are likely to progressively untether from fixed screens. This process is making room for more sophisticated motion-based interactions that also take greater consideration of body movements related to affective states, roleplaying, and social interactions with other players or non-playing characters. If the players' gaze is freed from having to stare at a fixed screen, the perception of one's own presence in the game will increasingly shift towards other sensory modalities, enabling players to perceive and interact with each other in a more natural and less mediated manner, thus allowing more intimate musical interactions. This would contribute to making video games an even more immersive and emotionally compelling experience: a medium increasingly useful for healthcare, education, and other fields beyond entertainment.

References

Alaoui, S.F., Caramiaux, B., Serrano, M., Bevilacqua, F.: Movement qualities as interaction modality. In: Proceedings of the Designing Interactive Systems Conference, pp. 761–769. ACM (2012)

Bernardo, F., Arner, N., Batchelor, P.: O Soli Mio : exploring millimeter wave radar for musical interaction. In: Erkut, C. (ed.) NIME'17 International Conference on New Interfaces for Musical Expression, pp. 283–286, Copenhagen (2017). http://homes.create.aau.dk/dano/nime17/papers/0054/paper0054.pdf

Bianchi-Berthouze, N.: Understanding the role of body movement in player engagement. Hum. Comp. Interact. **28**(1), 40–75 (2013). http://www.tandfonline.com/doi/abs/10.1080/07370024.2012.688468

Brown, D., Renney, N., Stark, A., Nash, C., Mitchell, T.: Leimu: gloveless music interaction using a wrist mounted leap motion. Proc. Int. Conf. New Interf. Mus. Expr. **16**, 300–304 (2016). http://www.nime.org/proceedings/2016/nime2016{_}paper0059.pdf

Burges, C.J.C.: A tutorial on support vector machines for pattern recognition. Data Min. Knowl. Disc. **2**, 1–43 (1998)

Camurri, A., Mazzarino, B., Ricchetti, M., Timmers, R., Volpe, G.: Multimodal analysis of expressive gesture in music and dance performances. In: Camurri, A., Volpe, G. (eds.) Gesture-based Communication in Human-Computer Interaction, pp. 20–39, vol. 2915, Springer, Berlin (2004). http://link.springer.com/10.1007/978-3-540-24598-8{_}3

Camurri, A., Coletta, P., Demurtas, M., Peri, M., Ricci, A., Sagoleo, R., Simonetti, M., Varni, G., Volpe, G.: A platform for real-time multimodal processing. In: Proceedings International Conference Sound and Music Computing, pp. 11–13 (2007). http://www.smc-conference.org/smc07/SMC07Proceedings/SMC07Paper59.pdf

Caramiaux, B., Françoise, J., Schnell, N., Bevilacqua, F.: Mapping through listening. Comput. Music J. **38**(3), 34–48 (2014)

CBS Interactive: Dance central for Xbox 360 reviews - Metacritic (2017a). http://www.metacritic.com/game/xbox-360/dance-central

CBS Interactive: Just dance for Wii reviews - Metacritic (2017b). http://www.metacritic.com/game/wii/just-dance

Champy, A.: Elements of motion: 3D sensors in intuitive game design. Analog. Dialogue. **41**(2), 11–14 (2007). http://www.analog.com/en/analog-dialogue/articles/3d-sensors-in-intuitive-game-design.html

Collins, K.: Game Sound : An Introduction to the History, Theory, and Practice of Video Game Music and Sound Design. MIT Press (2008). https://mitpress.mit.edu/books/game-sound

Cook, P.: Principles for designing computer music controllers. In: Proceedings of the International Conference on New Interfaces for Musical Expression 2001 (NIME 2001), pp. 3–6 (2001). http://www.nime.org/proceedings/2001/nime2001{_}003.pdf. http://dl.acm.org/citation.cfm?id=1085152.1085154

Die Gute Fabrik: Presskit — Johann Sebastian Joust (2014). http://www.jsjoust.com/presskit/

George, R.: Behind the Scenes of Zelda: Skyward Sword (2011). http://uk.ign.com/articles/2011/11/11/behind-the-scenes-of-zelda-skyward-sword

Glowinski, D., Dael, N., Camurri, A., Volpe, G., Mortillaro, M., Scherer, K.: Toward a minimal representation of affective gestures. IEEE Trans. Affect. Comput. **2**(2), 106–118 (2011). http://ieeexplore.ieee.org/lpdocs/epic03/wrapper.htm?arnumber=5740837. http://ieeexplore.ieee.org/document/5740837/

Godøy, R.I., Leman, M. (eds.): Musical Gestures: Sound, Movement, and Meaning. Routledge (2010)

Google ATAP: Project soli (2017). https://atap.google.com/soli/

Halston, T.: New PS4 camera vs old PS4 camera comparison what's the difference? (2016). https://www.psu.com/news/31081/New-PS4-camera-vs-old-PS4-camera-comparison---whats-the-difference

Holz, D.: Leap motion sets a course for VR (2014). http://blog.leapmotion.com/leap-motion-sets-a-course-for-vr/

Hunt, A.D., Wanderley, M.M., Paradis, M.: The importance of parameter mapping in electronic instrument design. In: Casey, C., Schneider, K., Hammond, E. (eds.) Proceedings of the 2002 Conference on New Instruments for Musical Expression (NIME-02), pp. 88–93, Dublin (2002). http://www.nime.org/proceedings/2002/nime2002{_}088.pdf

Jensenius, A.R.: To gesture or Not? An Analysis of Terminology in NIME Proceedings 2001–2013. In: Proceedings of the International Conference on New Interfaces for Musical Expression, pp. 217–220 (2014). http://www.nime.org/proceedings/2014/nime2014{_}351.pdf

Jensenius, A.R., Lyons, M.J.: A NIME Reader, Current Research in Systematic Musicology, vol. 3. Springer, Cham (2017). http://link.springer.com/10.1007/978-3-319-47214-0

Kuchera, B.: Nintendo Switch's World of Goo shows off system's Wii-style pointer controls (2017). https://www.polygon.com/2017/3/16/14947750/nintendo-switch-world-of-goo-controls-wii

Larboulette, C., Gibet, S.: A review of computable expressive descriptors of human motion. In: Proceedings of the 2nd International Workshop on Movement and Computing - MOCO '15. pp. 21–28. ACM Press, New York, NY (2015). http://dl.acm.org/citation.cfm?id=2790994.2790998. http://dl.acm.org/citation.cfm?doid=2790994.2790998

Lazzaro, N.: Why We Play Games: Four Keys to More Emotion Without Story. Technical Report, XEODesign Inc., Oakland, CA (2004)

Nintendo: How to calibrate the controllers (2017). http://en-americas-support.nintendo.com/app/answers/detail/a{_}id/22340/{-}/how-to-calibrate-the-controllers

Piana, S., Stagliano, A., Camurri, A., Odone, F.: A set of full-body movement features for' emotion recognition to help children affected by autism spectrum condition. In: IDGEI International Workshop, Crete (2013)

Piana, S., Stagliano, A., Odone, F., Camurri, A.: Adaptive body gesture representation for' automatic emotion recognition. ACM Trans. Interact. Intell. Syst. **6**(1), 1–31 (2016). http://dl.acm.org/citation.cfm?id=2896319.2818740

Rovan, J.B., Wanderley, M.M., Dubnov, S., Depalle, P.: Instrumental gestural mapping strategies as expressivity determinants in computer music performance. In: Proceedings of the AIMI International Workshop, pp. 68–73. Kansei-The Technology of Emotion Workshop (1997)

Serafin, S., Erkut, C., Kojs, J., Nilsson, N.C., Nordahl, R.: Virtual reality musical instruments: state of the art, design principles, and future directions. Comput. Music. J. **40**(3), 22–40 (2016). http://www.mitpressjournals.org/doi/10.1162/COMJ{_}a{_}00372

Sixense Entertainment: Razer hydra (2011). http://sixense.com/razerhydra-3

STMicroelectronics: Semiconductor solutions from STMicroelectronics selected by Nintendo for Nintendo switch (2017). http://www.st.com/content/st{_}com/en/about/media-center/press-item.html/t3934.html

Takahashi, D.: Nintendo Switch has high-tech Joy-Con controllers with motion detection-camera (2017). https://venturebeat.com/2017/01/12/nintendo-switch-has-high-tech-joy-con-controllers-with-motion-detection/

Trojaniello, D., Cereatti, A., Pelosin, E., Avanzino, L., Mirelman, A., Hausdorff, J.M., Della Croce, U.: Estimation of step-by-step spatio-temporal parameters of normal and impaired gait using shank-mounted magneto-inertial sensors: application to elderly, hemiparetic, parkinsonian and choreic gait. J. Neuroeng. Rehabil. **11**(1), 152 (2014). http://jneuroengrehab.biomedcentral.com/articles/10.1186/1743-0003-11-152. http://www.jneuroengrehab.com/content/11/1/152

Van Nort, D., Wanderley, M.M., Depalle, P.: Mapping control structures for sound synthesis: functional and topological perspectives. Comput. Music J. **38**(3), 6–22 (2014)

Visi, F.: Methods and technologies for the analysis and interactive use of body movements in instrumental music performance (2017). https://pearl.plymouth.ac.uk/handle/10026.1/8805?show=full

Visi, F., Schramm, R., Miranda, E.: Gesture in performance with traditional musical instruments and electronics. In: Proceedings of the 2014 International Workshop on Movement and Computing - MOCO '14. pp. 100–105. ACM Press, New York, NY (2014). http://dl.acm.org/citation.cfm?id=2618013. http://dl.acm.org/citation.cfm?doid=2617995.2618013

Visi, F., Caramiaux, B., Mcloughlin, M., Miranda, E.: A knowledge-based, data-driven method for action-sound mapping. In: NIME'17 International Conference on New Interfaces for Musical Expression (2017a)

Visi, F., Coorevits, E., Schramm, R., Miranda, E.R.: Musical instruments, body movement, space, and motion data: music as an emergent multimodal choreography. Hum. Technol. **13**(1), 58–81 (2017b). http://humantechnology.jyu.fi/archive/vol-13/issue-1/musical-instruments-body-movement-space-and-motion-data/@@display-file/fullPaper/Visi{_}Coorevits{_}Schramm{_}Reck-Miranda.pdf

Visi, F., Georgiou, T., Holland, S., Pinzone, O., Donaldson, G., Tetley, J.: Assessing the accuracy of an algorithm for the estimation of spatial gait parameters using inertial measurement units: application to healthy subject and hemiparetic stroke survivor. In: 4th International Conference on Movement Computing (2017c)

Wasenmuller, O., Stricker, D.: Comparison of kinect v1 and v2 depth images in terms of accuracy and precision. In: Lecture Notes in Computer Science (including subseries Lecture Notes in Artificial Intelligence and Lecture Notes in Bioinformatics). vol. 10117 LNCS, pp. 34–45. Springer, Cham (2017). http://link.springer.com/10.1007/978-3-319-54427-4{_}3

Wattanasoontorn, V., Boada, I., Garcia, R., Sbert, M.: Serious games for health. Entertain. Comput. **4**(4), 231–247 (2013). http://linkinghub.elsevier.com/retrieve/pii/S1875952113000153

Weichert, F., Bachmann, D., Rudak, B., Fisseler, D.: Analysis of the accuracy and robustness of the leap motion controller. Sensors. **13**(5), 6380–6393 (2013). http://www.ncbi.nlm.nih.gov/pubmed/23673678. http://www.pubmedcentral.nih.gov/articlerender.fcgi?artid=PMC3690061. http://www.mdpi.com/1424-8220/13/5/6380/

Wessel, D., Wright, M.: Problems and prospects for intimate musical control of computers. In: Poupyrev, I., Lyons, M.J., Fels, S.S., Blaine, T. (eds.) Proceedings of the 2001 International Conference on New Interfaces for Musical Expression (NIME-01), pp. 11–14, Seattle, WA (2001). http://www.nime.org/proceedings/2001/nime2001{_}011.pdf

Wöllner, C.: Body, Sound and Space in Music and Beyond: Multimodal Explorations. Routledge (2017)

Yang, L.: Knuckles cap sense overview (2017a). http://steamcommunity.com/sharedfiles/filedetails/?id=943495896

Yang, L.: Knuckles quick start (2017b). http://steamcommunity.com/sharedfiles/filedetails/?id=943406651

Zamborlin, B., Bevilacqua, F., Gillies, M., D'inverno, M.: Fluid gesture interaction design. ACM Trans. Interact. Intell. Syst. **3**(4), 1–30 (2014). http://dl.acm.org/citation.cfm?id=2543921. http://research.gold.ac.uk/9619/1/TiiS2013.pdf. http://dl.acm.org/citation.cfm?doid=2567808.2543921

Zappi, V., McPherson, A.: Dimensionality and appropriation in digital musical instrument design. In: Caramiaux, B., Tahiroglu, K., Fiebrink, R., Tanaka, A. (eds.) Proceedings of the International Conference on New Interfaces for Musical Expression, pp. 455–460. Goldsmiths, University of London, London (2014). http://www.nime.org/proceedings/2014/nime2014{_}409.pdf

Chapter 9
Repurposing Music According to Individual Preferences for Personalized Soundtracks

Duncan Williams

Introduction

This chapter briefly explores another futuristic vision for videogame soundtracking: the automated selection and manipulation of existing soundtrack material. An early feature in the *Grand Theft Auto* series allowed players to use their own libraries of music in the car radio. Naturally this might be problematic if the musical selections of the players become incongruous with the gameplay narrative at a given point, but the feature was nonetheless well regarded and successful due to the fact that individual musical preferences are so powerful (hence, why different radio stations with genre specific preferences can even exist in the first place in the 'real' world). Of course, the *Grand Theft Auto* example benefitted slightly in terms of immersion as, just like in the real-world, the player might have less control over the exact selection of music which appeared on their radio regardless of the necessary soundtracking.[1]

Imagine a situation where the player is somewhat responsible for the soundtrack selection, but the designer has then implemented a carefully selected series of protocols and effects in order to marry their selections with the narrative and to enhance both immersion and emotional congruence at any such point. How might this be possible? Emerging advances in the field of Music Information Retrieval (MIR) in combination with technology such as the AAC and BCMI approaches documented in earlier chapters of this book might be combined in a holistic system which allows users to specify some of their own soundtrack elements, but then allow specific affect matching processes to be synchronized with gameplay narrative. We will consider one such example which is now being commercially piloted by the Neurosky

[1] Motorhead's *Ace of Spades* causes the highest number of reported car crashes to occur in the UK when it is broadcast on commercial radio.

D. Williams (✉)
Digital Creativity Labs, University of York, York, UK
e-mail: duncan.williams@york.ac.uk

D. Williams, N. Lee (eds.), *Emotion in Video Game Soundtracking*, International Series on Computer Entertainment and Media Technology,
https://doi.org/10.1007/978-3-319-72272-6_9

group, who create simple BCI devices using electroencephalography, or EEG. This work has been heavily reported in the mainstream press, including by the BBC and Channel 5's *Gadgetshow* series, and has also been the subject of an academic journal paper (Eaton et al. 2015). EEG is a common choice for measuring electrical brain activity due to its non-invasive nature and relatively affordable and customisable hardware. However, interpreting meaningful information within EEG is a challenging task. Problems with noise inherent in complex signals that overlap into frequency or time-based ranges are commonly faced in BCI research. The quality of hardware components can make a significant difference in improving the signal-to-noise ratio and the ultimate success of mapping brainwaves to task related functions.

The Emotion Jukebox

Affective correlations to EEG have been suggested when mapping affective states to musical parameters (Daly et al. 2014) In other studies listeners have been asked to self-report emotions that are compared against such EEG readings (Daly et al. 2015). In this example, self-reported emotional states are used to evaluate a system for estimating arousal and valance from EEG by means of music selection from a real-time jukebox, with stimuli that have strong emotional connotations determined by a perceptual scaling analysis.

Clips with suggested emotional qualities based on crowd-sourced metadata were used in the study after being subjected to a perceptual scaling experiment in order to confirm their suggested emotional descriptors. Each clip is played back to a user over loudspeakers, generating a constant playback of musical excerpts in response to the affective state measured during the previous excerpt. Thus, a real-time prototype affective jukebox is realised. An experiment was conducted in order to validate the electroencephalogram (EEG) readings of the affective state reflected in, and interpreted via, the corresponding musical clips. To ascertain levels of valance and arousal electrodes were placed on the front of the scalp, over the prefrontal cortex an area that plays a significant role in emotion handling. Electrodes are positioned across points F3 and F4 (using the international 10–20 system). A subject's arousal can be derived from the ratio between alpha and beta activity. Strong alpha activity is known to indicate a relaxed state of mind, and this combined with increased activity in the beta band can indicate arousal; alertness in mental activity (Chanel et al. 2006). The balance of activation levels across the left and right hemispheres indicates a difference between a motivated approach or a more negative, withdrawal type of mental state, which is directly related to valence (Chanel et al. 2007) (Fig. 9.1).

Musical mood classification is a growing field in the realm of musical information retrieval, with various possibilities for stimulus selection including systems that utilise machine learning, crowd-sourcing, and acoustic analysis (Davies et al. 2011). In this case, the stimulus set was selected from music which had already been tagged

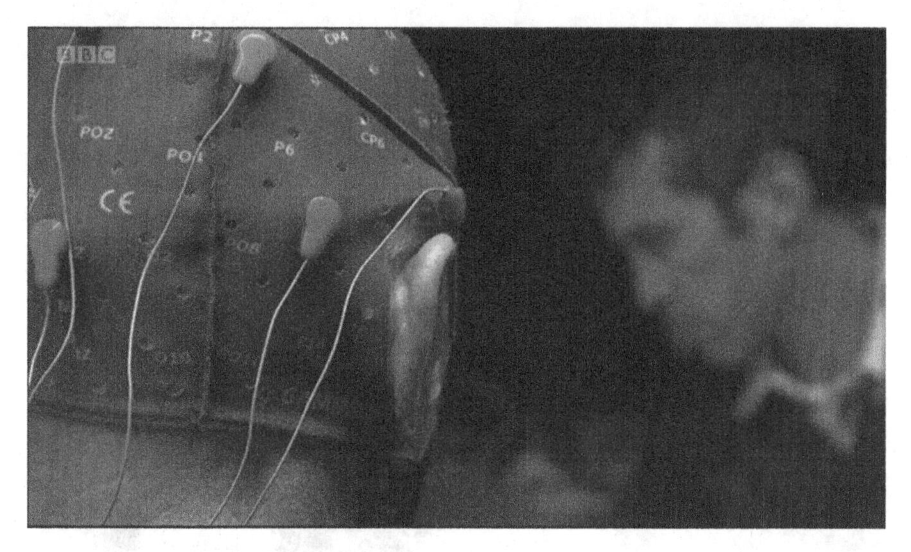

Fig. 9.1 A BBC news reporter demoing an 8-channel "Brain Jukebox" system for BBC News Spotlight, 6th July 2017, with the authors

with emotional descriptors by crowd-sourcing in the Stereomood project, an on-line radio service that aims to provide music that best suits users' mood and activities. The stimuli included material from a range of genres, with a variety of tempos and instrumentation. Material with tags that correspond to the affective adjectives shown in Fig. 9.2 was edited to 30-s clips and subjected to loudness equalisation in order to create the stimulus set shown in Table 9.1.

Statistical analysis of participant responses suggests that this approach can provide a feasible platform for further experimentation in future work. This could include using affective correlations to EEG measurements in order to control real-time systems for musical applications such as arrangement, re-composition, re-mixing, and generative composition via a neurofeedback mechanism which responds to listener affective states.

This pilot suggests that it is possible for a jukebox style affective music playback system to be controlled via EEG. Listener self-report confirmed that there was a good deal of corroboration between the EEG co-ordinates and individual affective state whilst engaging with the music selections.

Timbre and Spatialization as Parameters for Gameplay Cues

A second and third level of repurposing content might be achieved by timbral and spatial manipulation, as informed by biosensors or gameplay. Various approaches to spatialization are well documented in existing research with multichannel applications for composition (Malham 1998; Barrett 2002). Imagine, for example, the

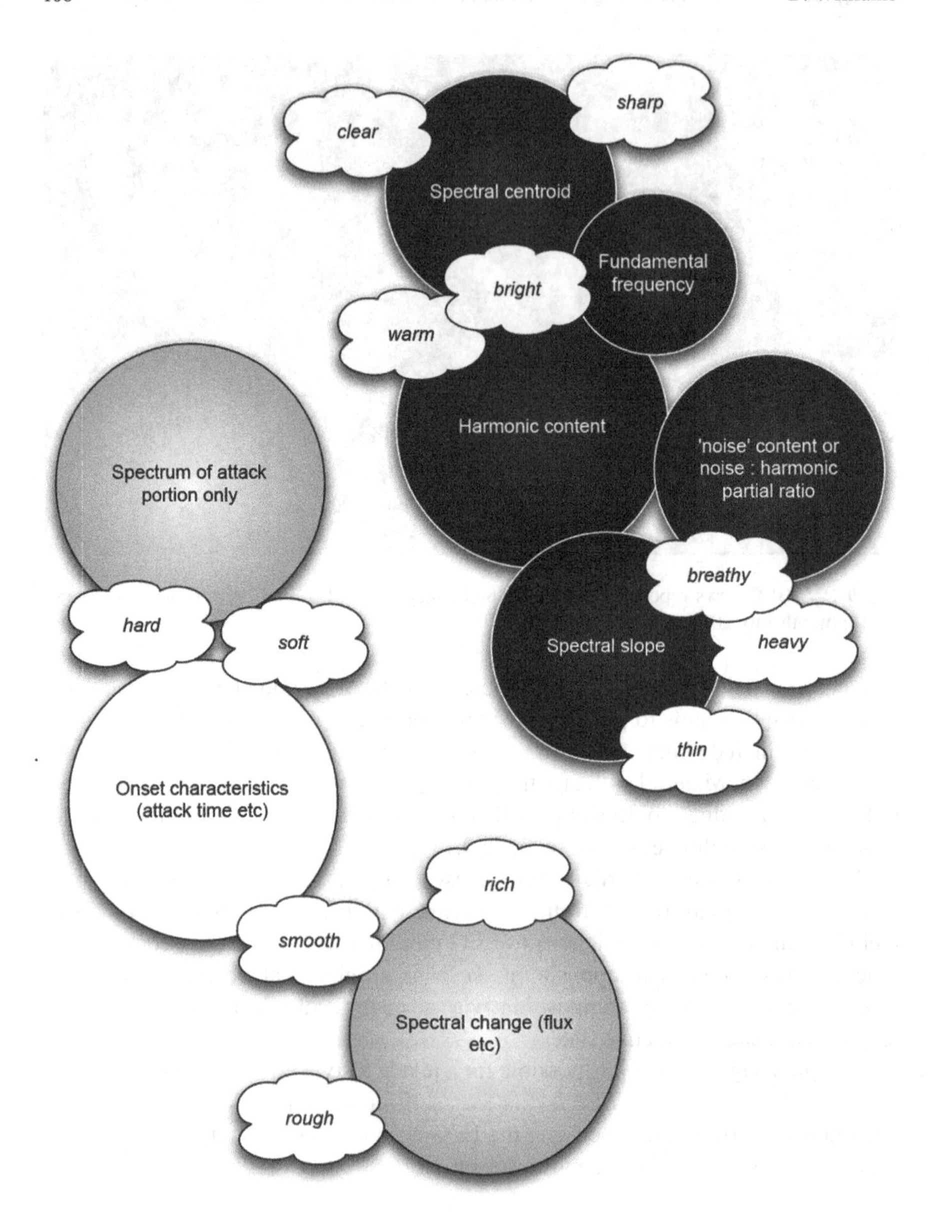

Fig. 9.2 Timbral attributes ('thought' bubbles) against acoustic correlates (circles). Spectral centroid, for example, is correlated with several timbral attributes. Acoustic correlates in black with white text are spectrally correlated only, shaded gray are temporally correlated, and white with black text are spectro-temporally correlated

Table 9.1 Stimulus set used in evaluation experiment

Stimulus number/Cartesian co-ordinate	Corresponding affective adjective	Musical stimulus (title, performer)
1. (a1, v1)	Tired	Dissociation EP, Gelatinous
2. (a2, v1)	Sad	One day I'll fly away, Keith Jarret & Charlie Haden
3. (a3,v1)	Miserable	Fade into you, Chelsea Burgin
4. (a4, v1)	Frustrated	What Kind of Girl, Kid Moxie
5. (a5, v1)	Angry	Sneak Chamber, Tsutchie & Force of Nature
6. (a6, v1)	Afraid	Perfect Drug, Nine Inch Nails
7. (a1, v2)	Relaxed	I'll take the road, Dave Reachill
8. (a2, v2)	Calm	Jung Greezy, Snake Oil
9. (a3, v2)	Content	Get Lucky, Daft Punk
10. (a4, v2)	Pleased	All around, Tahiti 80
11. (a5, v2)	Happy	Theo, Apples
12. (a6, v2)	Excited	Tropp'Cazz'Pa'Capa, Smania Uagliuns

Cartesian co-ordinates for arousal and valence are determined by EEG analysis. The corresponding affective adjective is then used to select a musical stimulus

audio morphing example given in Chap. 9, revised to incorporate specific timbral attributes. When asked to describe a sound, a player might use a combination of direct or metaphorical language (ASA 1960). This kind of descriptive timbral vocabulary is often a far cry from the quantitative approach used by sound designers to implement soundtracking (Barrett 2002; Berger 1964), presenting an obstacle for sound designers and signal processing engineers who seek to measure, and then manipulate such perceptual characteristics. To explore this more thoroughly, with a view to further work in the game design arena, we must consider timbre as a whole. Psychoacoustic research generally regards timbre as one of three perceptual attributes of sound along with loudness, and pitch. This list of perceptual attributes has been expanded to include perceived duration, spatial location, and reverberant environment (Cook 1999).

The American Standards Association (ASA), defines timbre as "… that attribute of sensation, in terms of which a listener can judge that two sounds having the same loudness and pitch are dissimilar" (ASA 1960). This definition can be considered a difficult starting point, as it does not define what timbre is, but rather what it is not. To illustrate this difficulty, we can consider unpitched or environmental sounds that would, according to the ASA definition, have no timbre. A more satisfactory definition is given by Pratt and Doak, as "the sensation on whereby a listener can judge that two sounds are dissimilar using other criteria than pitch, loudness or duration" (Pratt and Doak 1976).

Both loudness (Fletcher 1934) and pitch (Stevens 1937) have strong acoustic correlations that have been revealed and quantified by perceptual evaluation. Prior to the ASA definition of timbre, various timbral acoustic correlates were suggested,

Table 9.2 Showing various timbral attributes and suggested acoustic correlates

Timbral attribute	Acoustic correlate(s)
Breathiness (Johnson and Gounaropoulos 2006; Hillenbrand et al. 1994)	Spectral slope, periodicity
Brightness (Schubert and Wolfe 2006)	Spectral centroid, fundamental frequency
Clearness, Clarity (Disley et al. 2006)	Harmonic amplitude, spectral slope, harmonic ratio
Richness (Stevens and Harris 1962)	Fluctuation of energy between adjacent partials
Roughness (Zwicker 1961; Stevens and Harris 1962; Terhardt 1974; Aures 1985)	Aspers, critical bands and partials above 6th Harmonic
Sharpness (Von Bismarck 1974a; Eberhard Zwicker and Fastl 1999)	Spectral centroid, function of critical bandrate
Smoothness (Stevens and Harris 1962)	Amplitude envelope, also antonym of roughness, hardness, synonym of softness
Warmth (Pratt and Doak 1976)	Low frequency content, spectral slope, ratio of first harmonic to other spectral content

including Harmonicity (Seashore 1934), number, distribution, and relative intensity of partials (spectral characteristics). Schouten (1968) subsequently expanded upon this list of acoustic correlates to include:

- The range between tonal and noiselike character.
- The spectral envelope.
- The time envelope in terms of rise, duration, and decay.
- The changes both of spectral envelope (formant-glide) and fundamental frequency (micro-intonation).
- The prefix, an onset of a sound quite dissimilar to the ensuing lasting vibration.

A Note About Tone Colour and Sound Quality

Tone colour and sound quality are terms that might be used synonymously with timbre. However, tone colour can imply that the spectral properties of the sound are solely responsible for its timbre, contradictory to research specifying the importance of temporal acoustic correlates with relation to the perception of timbre (Iverson and Krumhansl 1993). Moreover, sound quality is most often found associated with playback, stereophony, multichannel sound, and loudspeaker performance measurement.

Various different approaches to the research of timbre, and subsequent timbral paradigms, inform the timbral descriptors used in existing research. Table 9.2 shows various timbral attributes against suggested acoustic correlates.

On closer examination of the timbral attributes shown in Table 9.1, various discrete categories can be found, wherin the timbral descriptors are either informed by the physical properties of the instrument, the performance style employed by the

Table 9.3 A summary of existing perceptual measures (excluding loudness and pitch), showing their corresponding units where known

Perceptual measure	Unit
Subjective duration	Dura (Zwicker and Fastl 1999; Hillenbrand et al. 2000)
Sharpness	Acum (Von Bismarck 1974a; Zwicker and Fastl 1999)
Brightness	n/a (Schubert and Wolfe 2006)
Roughness	Asper (Stevens and Harris 1962; Terhardt 1974; Aures 1985)
Fluctuation Strength	Vacil (Terhardt 1974; Zwicker and Fastl 1999)
Tonalness	n/a (Zwicker and Fastl 1999)
Sensory pleasantness	n/a (Zwicker and Fastl 1999)

musician, or, other sensual analogies (visual, tactile, gestatory, etc). Whilst not mutually exclusive, the categories do provide an idea of how these metaphors and analogies can be useful to the game sound designer as direct timbral attributes.

Boldness, smoothness, hardness and so on could be derived from the performance of the soundtrack itself. Descriptors such as breathiness in Table 9.2, or similarly punchy, gentle, and so on are likely to be extracted from the performance of the speech artist (Gabrielsson 1996).

Clearness, and warmth can similarly be derived from sensual analogies (visual in the case of clear, and with temperature in the case of warmth). Descriptors such as metallic and hollow could be extrapolated by the listener from the physical properties of the game element producing the sound—for example, a sword might sound metallic (Wedin and Goude 1972; Lutfi 2001; Nykanen et al. 2009).

Figure 9.1 highlights some overlap amongst these acoustic correlates, based on their allotted timbral attributes.

The range of perceptual descriptors which are included in timbre studies are reflected in their acoustic correlates and the lack of a unidirectional scale suitable for use in existing game sound engines (Pratt and Doak 1976; Krumhansl 1989; Mcadams 1999; Johnson and Gounaropoulos 2006). Moreover, there are limited efforts linking timbral characteristics to emotion.

Bresin (Bresin and Friberg 2000) and Gabrielsson amongst others have carried out several studies as to the acoustic correlates and timbral descriptors which contribute to the sense of emotion perceived in musical performances. The basic experimental procedure is to instruct professional musicians to perform on their chosen instrument (violin, guitar, voice, and flute in Gabrielsson's case), with the intention of communicating specific emotional descriptors to the listening panel. Listeners then rate the emotion, allowing for a detailed analysis of any acoustic correlates, including both dynamics and spectral features.

A number of perceptual dimensions have already been quantified, as shown in Table 9.3. Note that some of these dimensions can be considered physically independent (sharpness/brightness, roughness, fluctuation strength) whilst others include some physical overlap (tonalness, pleasantness).

Brightness is often shown to be highly correlated with spectral centroid, or a combination of spectral centroid and fundamental frequency (Schubert and Wolfe 2006). It is worth noting that an increase of fundamental frequency would probably, by definition, increase the spectral centroid.

The work of Gabrielson mentioned earlier (Gabrielsson 1996) is closest to the spirit of emotionally-driven game soundtracking. In order to establish the acoustic correlates of performance timbres, Gabrielson asks listeners to describe short melodies in terms of their emotional impact and performance. The stimulus set is then analysed acoustically to determine which spectral and temporal characteristics could be shown to contribute to the elicitated attributes, therby arriving at a similar model to the 'measurement' approaches. For example, two passages comprised of the same notes, but played in a different rhythm, could be acoustically the same (in terms of pitch, and loudness), but differentiable by their performance, equating to a change in timbre according to the ASA definition of timbre. Examples of this type of timbral attribute could include staccato (detached, or rather, detached-ness), espressivo (expressively, or expressive-ness), or giocoso (playful, or playful-ness).

This steps involved in the development of a game-driven procedure would be as follows

- Determine acoustic correlation(s)
- Manipulate acoustic correlation(s) by digital signal processing
- Measure emotional change in stimuli by perceptual experiment (ask players to rate emotional change)
- Label direction and type of emotional change, implement controller in game engine

Together with the repurposing example illustrated earlier in this chapter, we might then see adaptively driven game engines which manipulate particular timbral attributes of an existing soundtrack, repurposing it according to the best emotional intent to match the gameplay or narrative.

The nomenclature is mostly universal, although a large number of labels and descriptors have not been acoustically quantified as yet. Work to reduce the range of descriptors, perhaps down to those which are acoustically independent, and a subsequent set of those with acoustic overlap would be useful. Attributes with acoustic overlap, or indeed attributes which appear to have contradictory acoustic correlates would also require the ratio between their acoustic correlates to be quantified for further work if they are to be included in the proposed emotion-driven audio engine, allowing for targeted manipulation, quantification, and verbal description of specific timbral attributes by digital signal processing. In the future, the adapted paradigm could be 'fleshed out' to include as many timbral attributes as possible, which could be controlled by gameplay for example as part of a discrete signal processing routine in the game engine.

References

ASA: American Standard Acoustical Terminology, Definition 12.9, Timbre. American Standards Association, New York (1960)

Aures, W.: A procedure for calculating auditory roughness. Acust. **58**, 268–281 (1985)

Barrett, N.: Spatio-musical composition strategies. Organ. Sound. **7**, 313–323 (2002)

Berger, K.W.: Some factors in the recognition of timbre. J. Acoust. Soc. Am. **36**, 1888 (1964). https://doi.org/10.1121/1.1919287

Bresin, R., Friberg, A.: Emotional coloring of computer-controlled music performances. Comput. Music. J. **24**, 44–63 (2000)

Chanel, G., Kronegg, J., Grandjean, D., Pun, T.: Emotion assessment: arousal evaluation using eeg's and peripheral physiological signals. In: Gunsel, B., Jain, A.K., Tekalp, A.M., Sankur, B. (eds.) Multimedia Content Representation, Classification and Security, pp. 530–537. Springer LNCS (2006)

Chanel, G., Ansari-Asl, K., Pun, T.: Valence-arousal evaluation using physiological signals in an emotion recall paradigm. In: Systems, Man and Cybernetics, 2007. ISIC. IEEE International Conference on, pp. 2662–2667 (2007). https://doi.org/10.1109/ICSMC.2007.4413638

Cook, P.R. (ed.): Experimental design in psychoacoustic research. In: Music, Cognition, and Computerized Sound: An Introduction to Psychoacoustics, pp. 299–320. MIT Press, Cambridge, MA (1999)

Daly, I., Malik, A., Weaver, J., Hwang, F., Nasuto, S.J., Williams, D., Kirke, A., Miranda, E.: Identifying music-induced emotions from EEG for use in brain-computer music interfacing. In: IEEE, pp. 923–929 (2015), https://doi.org/10.1109/ACII.2015.7344685

Daly, I., Malik, A., Hwang, F., Roesch, E., Weaver, J., Kirke, A., Williams, D., Miranda, E., Nasuto, S.J.: Neural correlates of emotional responses to music: an EEG study. Neurosci. Lett. **573**, 52–57 (2014)

Davies, S., Allen, P., Mann, M., Cox, T.J.: Musical moods: a mass participation experiment for affective classification of music. In: ISMIR, pp. 741–746 (2011)

Disley, A.C., Howard, D.M., Hunt, A.D.: Timbral description of musical instruments. In: Proceedings of the 9th International Conference on Music Perception and Cognition, Bologna (2006)

Eaton, J., Williams, D., Miranda, E.: The space between us: evaluating a multi-user affective brain-computer music interface. Brain Comput. Interf. **2**, 103–116 (2015)

Fletcher, H.: Loudness, pitch and the timbre of musical tones and their relation to the intensity, the frequency and the overtone structure. J. Acoust. Soc. Am. **6**, 59–69 (1934). https://doi.org/10.1121/1.1915704

Gabrielsson, A.: Acoustic correlates of emotionally expressive music. J. Acoust. Soc. Am. **100**, 2778 (1996). https://doi.org/10.1121/1.416426

Hillenbrand, J.M., Clark, M.J., Houde, R.A.: Some effects of duration on vowel recognition. J. Acoust. Soc. Am. **108**, 3013 (2000). https://doi.org/10.1121/1.1323463

Hillenbrand, J., Cleveland, R.A., Erickson, R.L.: Acoustic correlates of breathy vocal quality. J. Speech Hear. Res. **37**, 769–778 (1994)

Iverson, P., Krumhansl, C.L.: Isolating the dynamic attributes of musical timbre[a]. J. Acoust. Soc. Am. **94**, 2595–2603 (1993). https://doi.org/10.1121/1.407371

Johnson, C.G., Gounaropoulos, A.: Timbre interfaces using adjectives and adverbs. In: New Interfaces for Musical Expression, pp. 101–102. IRCAM, Paris (2006)

Krumhansl, C.L.: Why is musical timbre so hard to understand? In: Structure and Perception of Electroacoustic Sound and Music, pp. 43–53. Elsevier, Amsterdam (1989)

Lutfi, R.A.: Auditory detection of hollowness. J. Acoust. Soc. Am. **110**, 1010 (2001). https://doi.org/10.1121/1.1385903

Malham, D.G.: Tutorial article: approaches to spatialisation. Organ. Sound. **3**, 167–177 (1998)
Mcadams, S.: Perspectives on the contribution of timbre to musical structure. Comput. Music. J. **23**, 85–102 (1999)
Nykanen, A., Johansson, O., Lundberg, J., Berg, J.: Modelling perceptual dimensions of saxophone sounds. Acust. United Acta Acust. **95**, 539–549 (2009)
Pratt, R., Doak, P.: A subjective rating scale for timbre. J. Sound Vib. **45**, 317–328 (1976). https://doi.org/10.1016/0022-460X(76)90391-6
Schouten, J.F.: The perception of timbre. In: Reports of the 6th International Congress on Acoustics, GP-6-2, pp.35–44, Tokyo (1968)
Schubert, E., Wolfe, J.: Does timbral brightness scale with frequency and spectral centroid? Acta Acust. United Acust. **92**, 820–825 (2006)
Seashore, C.E.: The timbre of orchestral instruments. J. Acoust. Soc. Am. **6**, 55 (1934). https://doi.org/10.1121/1.1915695
Stevens, S.S.: A scale for the measurement of the psychological magnitude pitch. J. Acoust. Soc. Am. **8**, 185 (1937). https://doi.org/10.1121/1.1915893
Stevens, S.S., Harris, J.R.: The scaling of subjective roughness and smoothness. J. Exp. Psychol. **64**, 489–494 (1962). https://doi.org/10.1037/h0042621
Terhardt, E.: On the perceptions of periodic sound fluctuation (roughness). Acust. **30**, 201 (1974)
Von Bismarck, G.: Sharpness as an attribute of the timbre of steady sounds. Acustica. **30**, 172 (1974)
Wedin, L., Goude, G.: Dimension analysis of the perception of instrumental timbre. Scand. J. Psychol. **13**, 228–240 (1972). https://doi.org/10.1111/j.1467-9450.1972.tb00071.x
Zwicker, E.: Subdivision of the audible frequency range into critical bands (Frequenzgruppen). J. Acoust. Soc. Am. **33**, 248 (1961)
Zwicker, E., Fastl, H.: Psychoacoustics: Facts and Models, 2nd updated edn. Springer, New York (1999)

Chapter 10
Sounding the Story: Music in Videogame Cutscenes

Giles Hooper

Definitions

> … many of the most memorable emotional moments I as a gamer come to remember have happened in front of my eyes when I had lost the control of the action through a triggered cut-scene of varying length … (Perron 2016)

While most people assume they are talking about the same thing, it is salutary to note that the 'videogame cutscene' does not in fact circumscribe a discrete and objectively-identifiable category of in-game event. Nevertheless, a commonly encountered definition is that cutscenes—also referred to as 'cinematics' or 'in game movies'—are pre-determined/pre-scripted audio-visual sequences which do not involve direct player intervention; or which, as Sicart puts it, are 'devoid of any procedural agency' (2012, p. 120). The latter is typical of the way in which cutscenes are normally defined apophatically (i.e. as what they are not). A more positive definition is offered by O'Grady, who describes the cutscene as 'a spatiotemporal unit, often freighted with narrative, dramatic, or spectacular significance' (2013, p. 108).

Videogame historians and scholars normally consider the first cutscenes to be the 'intermissions' which appear in the original 1980 arcade version of *Pac Man*. For example, in citing an instance of non-dynamic non-diegetic audio, Collins refers to 'sound that is part of an underscore that is unaffected by a player's movements, such as the cut-scene after the second level of Pac-Man, in which gameplay is stopped so a quick "film" can play' (2007, p. 212). These were little more than 'humorous' interludes. Nevertheless, they appear to satisfy the baseline criteria—albeit one

G. Hooper (✉)
Department of Music, University of Liverpool, Liverpool, UK
e-mail: giles.hooper@liverpool.ac.uk

D. Williams, N. Lee (eds.), *Emotion in Video Game Soundtracking*, International Series on Computer Entertainment and Media Technology,
https://doi.org/10.1007/978-3-319-72272-6_10

might prefer to think of them as 'proto-cutscenes'—insofar as they involve no direct intervention on the part of the player; and even, in a most limited way, serve to imply some notion of anthropomorphic 'characterisation'.[1]

Types

While Collins describes a cutscene as 'a movie that plays in a game, in which the player does not control any actions' (2007, p. 212), there are nominal cutscenes which nevertheless permit, or even require, some limited player intervention. Accordingly, Fritsch adopts a more flexible conception of cutscenes, defining them as '… intermission[s] of gameplay in which the player *usually* cannot or can only *slightly* influence events on screen' (2013, p. 13; my emphasis). For example, some cutscenes (usually in first-person) allow the player to manipulate the camera within a narrow angle-range, as is the case with the (in)famous 'car journey' scene which occurs near the beginning of *Call of Duty 4: Modern Warfare* or the extended introductory 'cart journey' sequence in *Skyrim*. The *BioShock* series is notable for utilising this kind of 'scripted event', eschewing purely cinematic cutscenes and almost always allowing the player to at least retain camera-control and so look around (even if otherwise inhibited by some form of in-game 'conceit', such as the bathysphere descent in the first game of the series).

In other instances, particularly in Japanese role-playing games (JRPGs), such as the *Tales* series, players are often required to press a button or key, repeatedly, to advance a pre-scripted dialogue via a series of text-box call-outs (this remains true even of latest generation releases such as *Tales of Berseria* [2016]); this is another 'conceit' mechanic (a way of controlling the presentation of an extended dialogue in the absence of voice-acted delivery).[2] Weiss (2003) instead proposes 'dialogue-scene', indicating 'anything which, although it may be stylistically similar to a cinematic cut-scene, requires the minimal action of the player pressing the button to scroll through it'. Mechanical 'dialogue-scenes' such as these are nevertheless to be distinguished from those sequences in which a player must decide between dialogue-choices as part of an active 'in-gameplay' conversation with a non-player character (NPC).[3]

---

[1] Many videogames include short pre-rendered 'stock' scenes, which accompany a normally infrequent but repeatable in-game action or event, such as the standard post-battle RPG 'victory pose', or those which are cued when a player chooses to rest at a bonfire in *Dark Souls*. Despite temporarily closing-off player intervention, it remains a moot point as to whether these should be considered cutscenes. Because of its presumed expository nature, a cutscene is normally a unique event within a single playthrough of the game.

[2] It is also a way of significantly reducing development cost.

[3] 'Gameplay' is normally, if questionably, understood along the lines set out by Salen and Zimmerman: 'the formalized interaction that occurs when players follow the rules of a game and experience its system through play' (2003, p. 303).

Finally, there are ‘cutscenes’ which play-out in a different way depending on a choice which is made by the player during the sequence, such as the time-limited ‘paragon’ or ‘renegade’ options that sometimes flash on screen during *Mass Effect 3* (and which can have a significant impact on the remainder of the game—the choice precipitating the death of Mordin Solus representing perhaps the most profound, and moving, example); or there are otherwise pre-determined ‘success or failure’ sequences which require a button to be pressed at a specific moment, such as in *Resident Evil 4* (see also Cheng 2007). Such scenes bear a family-resemblance to the short-lived laser-disc games of the 1980s. For some, such sequences are *de facto* cutscenes which incorporate an interactive element; for others, these are *de jure* not cutscenes, precisely because they incorporate the same. The argument quickly becomes circular; and, rather like the Sorites paradox, it raises the question: how interactive must a sequence be (or how many button-presses must be required) before it ceases to be a cutscene?

The preceding discussion may appear an exercise in semantics (or pedantry); yet, what is intended (i.e. included/excluded) by the designation ‘cutscene’ obviously has significant import in respect of subsequent theorisation. With that in mind, it is possible to derive a provisional typology.

Cutscene	
Cinematic	No camera control; no interaction
Viewpoint	Camera control; no interaction
Fixed dialogue	Player required to manually advance pre-determined dialogue
Hybrid	
Branching	Player-decision results in alternative continuation of cutscene
Quick time event	Player required to make indicated input at indicated point
Others	
Stock animation	Short repeatable sequence following a repeatable in-game event
Location cutaway	Short camera-pan introducing/showing an area
In-game dialogue	Player makes active dialogue choice(s)

The liminal instances test the attributions applied to them, and assist in avoiding the circularity of ‘confirmation bias’ (whereby cutscenes are delimited a priori by the function one already wishes to attribute to them or the criticism one wishes to make of them). For the purposes of this chapter, discussion relates mainly to those types referred to as ‘cinematic’ and ‘viewpoint’.

Implementation

In terms of technical implementation, there are two principal ways of realising cutscenes: ‘pre-rendered’, where the sequence is an independent audio-visual entity (which could include FMV); and ‘real-time’, where the sequence is rendered by the game-engine and (usually modified) in-game assets. Historically, the former afforded more sophisticated graphical presentation, since it allowed the use of externally-produced CGI and other techniques, which saw cutscenes approach the quality of contemporaneous animated films. However, such sequences require significant time, resource, and memory-usage.

In addition, the transitions to and from such cutscenes, despite increasing ingenuity in their handling, could still prove abrasive; and the contrast in the quality of graphical presentation (between the cutscene and the remainder of the game) was often stark. Writing at a time when such incongruity was at its most obvious—the CD-ROM era—Howells argued that 'As the game strives to make gamers believe the imaginary, computer generated—and often blocky and pixelated – game world, the transition to full-motion video reminds gamers that this is, in fact, not real, breaking the suspension of disbelief' (2002, p. 113). This was especially apparent, for example, in the trio of PS1 *Final Fantasy* games (*VII* through *IX*).[4] As Rouse—an industry-insider—noted at the time: 'What concerns me most are cut-scenes which don't graphically fit in any way with the game they supplement, movies which seem to have been filmed in an entirely different universe from the one that the player encounters in the game itself. Surprisingly enough, this is the norm for our industry' (1998, p. 7). Explanations for the inclusion of such scenes can therefore tend toward the cynical, viewing them as either the creations of frustrated producers or art-directors who would prefer to be working in film; or marketing ploys on the part of the developers (consider how many videogame advertisements include the small-print caveat 'not actual in-game footage'). However, I would propose an alternative pair of rationales: on the one hand, cutscenes became the new 'reward' mechanism (as the pursuit of 'high scores' waned in significance); on the other hand, they served to frame the videogame as more than simply a game (as videogaming became an increasingly adult pursuit).[5]

Moreover, because many games, especially role-playing games (RPGs), allow significant character customisation, purely pre-rendered cutscenes cannot be used—without disruptive or even unintentionally humorous 'continuity errors'—if they are to feature the player avatar or any player-customised NPC. Thon (2016) highlights a number of such 'inconsistencies', in *Halo*, *Dragon Age: Origins*, and *The Witcher 2*. However, he appears to believe these to be due to a combination of representational convention and a 'principle-of-charity' tolerance on the part of players who will attribute it to an oversight on the part of the developers. In the main, such discrepancies were more likely the product of a necessarily parsimonious management of in-game assets as well as a desire for the cinematic potential (and staged 'punch')

[4]The contrast was most obvious in the case of CD-ROM consoles because, in the transition from the former cartridge-based systems, external memory capacity—which permitted the storage and streaming of high resolution CGI—increased by a significantly higher factor than did CPU speed/power, RAM size and real-time graphical capabilities.

[5]Most surveys suggest that approximately 70–75% of videogame-players are adults, and that the average age of a videogame-player is early-mid 30s. The marketing of the Sony PS1 played a significant role in transforming the perception of videogaming from an isolated adolescent pursuit to a social adult pursuit. Many triple-A releases carry an adult-rating.

of pre-determined scenes overriding the continuity fidelity afforded by real-time rendering. Nevertheless, fully pre-rendered cutscenes have been utilised less frequently since the early-mid 2000s.[6]

Consequently, some form of 'real-time' rendering is now typically used, especially as technological advance has allowed the realisation of game-engine cutscenes which are closer in quality to those of pre-rendered CGI, while simultaneously permitting a more sophisticated and fluid integration with that which precedes and follows (see also King and Krzywinska 2002; Cheng 2007). They may also, as noted above, allow a player to manipulate the camera-angle during a cutscene (something which has obvious ramification for VR, to which I return in the latter part of this chapter). In practice, while generated by the game-engine, they normally require an additional portfolio of bespoke assets (e.g. higher-resolution character models, modelled lighting effects, custom animations, specific code-scripting etc). This can significantly close, but does not strictly eliminate, the discrepancy in graphical fidelity between cutscene and gameplay environment.[7]

Perspectives

Most contemporary videogames incorporate at least some form of cutscene, as did a significant proportion of earlier videogames (depending on the stricture of the definition). However, within the field of academic videogame studies, there is less engagement with cutscenes, and very little in respect of music in cutscenes; or, rather, there is very little *explicit* engagement. When cutscenes are discussed, they are typically enjoined as a secondary or supporting element in a wider argument.

Stories

One possible explanation for this (see also King and Krzywinska 2002; Juul 2010), is that cutscenes appear to negate what many believe to be the primary, and differentiating, fact of videogames: their ludic and supposedly interactive nature. Eskelinen once put it rather bluntly: '… stories are just uninteresting ornaments or gift-wrappings to games, and laying any emphasis on studying these kinds of marketing tools is just a waste of time and energy' (2001). I dispute this claim, since, in many videogames, 'stories' are central to contextualising and motivating a

[6] Part of the marketing and publicity around *Uncharted 4* (PS4 2016) involved the fact that all of the cutscenes were rendered in real-time via the in-game engine. Such claims have also provoked controversy in the past: for example, the famous *Killzone 2* trailer, which the developer claimed was 'in-game', but was later revealed to be a 'target render'.

[7] An additional advantage of 'real time' rendering is that it tends to produce better results in respect of scaling the output resolution to the technical specification of the system being used.

continued engagement with the ludic/ergodic mechanics (and, as we have seen, even *Pac Man*, in however rudimentary a fashion, sought to engender some vague sense of identification). While stories are not the same as cutscenes, since it is possible to convey a story without them, Eskelinen's (strong ludological) perspective nevertheless explains why cutscenes—at least extended or 'non-skippable' cutscenes—are often viewed negatively by players and critics alike.[8]

As long ago as 2002, in another forthright contribution—now a founding text of videogame scholarship—Newman adopted a contrary, if equally radical, line, whereby he proposed that 'Quite simply, videogames are not interactive, or even ergodic'. His point, sometimes misunderstood, was that a videogame *in toto* comprises a diverse range of highly structured and segmented experiences, only some types of which—even if normally accounting for the bulk of game time—constitute the active player intervention/input sequences which nevertheless tend to inform both the ludological and vernacular conception of 'gameplay'. It is interesting to note that even the Wikipedia entry for 'Cutscene' opens with 'A cutscene or event scene … is a sequence in a video game that is not interactive, breaking up the gameplay'; and, similarly, Summers observes that 'Cutscenes may also interrupt the middle of levels at particularly significant moments' (2016, p. 21). This is not to suggest that 'breaking up' or 'interrupt' are here intended pejoratively—in our everyday lives, we often speak of wanting to 'break up the boredom'. Rather, it is to highlight, in almost Derridean fashion, the implicit separation and hierarchy thus established, whereby cutscenes are the supplemental things which punctuate the primary thing—the 'gameplay', the 'level'.

Klevjer (2002) is one of few not only explicitly to have engaged with cutscenes, but also to have suggested a different conception of their role and significance. His literal 'defence of cutscenes' was made at the height of the 'ludological versus narratological' debate, which would then become central to gaming-studies for a decade; and persists, even if in modified form (see also Moberley 2013).[9] While acknowledging the importance of the ludic and ergodic dimensions, Klevjer was concerned that what he termed 'radical ludology' would become conflated with the possibility of 'pure' gaming, a kind of '*an sich*' abstraction. He argued that a 'cutscene does not cut off gameplay', but rather is 'an integral part of the configurative experience' (2002, p. 195). In other words, instead of conceiving of a videogame as the 'gameplay', framed and punctuated by secondary elements, the concept of the 'configurative experience' acknowledges an holistic notion of 'a videogame' as informed by the full range of incorporated elements: not only 'gameplay', but also cutscenes, title screens, menu screens, end-credits, and similar (one might also

[8] This negative reception partly informed the greater use, by some developers, of QTEs (Quick Time Events), in which a de facto cutscene requires the player to press a given button at a given time. However, if anything, QTEs were received even less favourably, since those not disposed to cutscenes saw little enhancement in being asked to press a single button when prompted; while those who were disposed to cutscenes felt that the emotional impact was lessened or the purpose of the cutscene diluted for the same reason.

[9] Several of the defining texts concerned with narrative and cutscenes date from the early 2000s—some time ago—precisely because they were central to the incipient narratology-ludology 'debate'.

include marketing, publicity, intertextual references, official guides, online help-sites and walkthroughs, community threads, and so on—since each of these may inform a player's interaction with, and experience of, a videogame).[10]

Interactions

Much therefore depends on whether cutscenes are considered intrinsic or extrinsic to the 'gameplay'. The simple (or simplistic) syllogism runs thus: videogame gameplay is interactive; cutscenes are not interactive; therefore, cutscenes are not a part of videogame gameplay. However, in a version of the *petitio principii* fallacy, this assumes a notion of interactivity predicated on a kinetic-cybernetic feedback loop within a rules-based system—one which also presupposes, by analogical default, a passivity when engaging with a book or a film. Yet this is despite the fact that post-Barthesian theory has long emphasised the active and constructive role of the reader/viewer. A similar point is made by O'Grady (2013), who notes that both phenomenology and cognitive science have convincingly proposed that to make meaning of what one is seeing and hearing is itself an 'active' and 'embodied' (affective) process.

In other words, the interactivity predicated of games tends to derive from two related emphases: firstly, the ergodic, rules-based, goal-oriented nature of games, in which the player is required continuously to (re)configure the 'text' with which they are engaging; secondly, the iterative 'causal event loop' of observation, cognitive process, reaction, and kinetic input. However, as Cheng notes, drawing on Rehak, the ludological conception of interactivity is thereby narrowed, only occurring 'when the player has his hands on the controller, in other words, as part of the cybernetic, or homeostatic feedback loop when the player engages in "some motor action via an interface"' (2007, p. 16). His productive move is to shift the emphasis from (mechanical/kinetic) interaction to 'agency'—the affective response a player feels in seeing the consequence of their input; and then, via Klevjer, to note how cutscenes play a central role in creating the contextualising/motivational framework within which those player actions become meaningful (or at least more meaningful) and thereby affecting.

Excepting an opening cutscene, later sequences carry, within the player's mind, the potential resonance or 'trace' of preceding actions or choices. As Newman observes, 'Narrative sequences can necessitate their own level of interactivity, requiring a certain degree of commitment from the player. It is true that this interactivity differs in quality from that of play activity itself, but it is a grave error to characterize it simply and in a contradictory fashion as an incitement to 'passivity'' (2004, p. 97). For example, if a player opts, in *The Witcher 3*, for one of the choices

[10] Summers (2016) likewise points to 'those [musical] cues that are usually ignored in analyses of game music (such as cues for developer logos, or menu music), but that are nevertheless significant components of the game as a musical source' (p. 15).

which results in their discovering, via a later cutscene, that the NPC Keira Metz was subsequently burned at the stake, it is implausible to argue that they are only processing the audio-visual modal duality presented to them; or that the emotional response is simply the same as that which might be provoked by a similar scene in a film such as *The Name of the Rose*. As Fahlenbrach (2016) notes: 'Typically, narrative emotions rely on varying degrees of closeness to and emotional engagement with characters – both by means of imaginative perspective-taking and by interacting with them in more or less complex narrative scenarios'. This is similar to the difference between 'induced emotions' and 'observed emotions' (Juslin 2001). Baseline empathetic reaction may be the same, as might elements of visceral shock or disgust; but there is a crucial difference: this was not a pre-determined outcome (however many times one watches *Star Wars*, Obi-Wan will always 'die' at the hands of Darth Vader; but, in *The Witcher 3*, Keira Metz may or may not die). The player's own 'agency' is now writ large; and so the putative hard-boundary binary between 'gameplay'—in which the player made the requisite choice—and 'cutscene'—in which the consequence of the player's choice is subsequently represented—is rendered tenuous.[11]

Excursus: Immersion

As is apparent—if implicitly and indirectly—from the preceding sections, the claim that cutscenes are 'immersion-breaking' underpins much of the criticism directed at them. This depends of course on how one is conceiving or defining immersion. From the ludological perspective, it is the absence (or prevention) of interaction that damages immersion (*qua* ergodic engagement); conversely, from the narratological perspective, cutscenes can enhance immersion (*qua* imaginative absorption).

One can be immersed in a game of *Tetris*; yet, excepting a theoretically extreme form of radical ludology, this is not the same as being immersed in a game of *Deus Ex*.[12] Although a player may, in one sense, be equally immersed (cognitively and kinetically absorbed) in either, it is unlikely that they imagine the defined *Tetris* screen-space to represent a world in which they empathetically identify with coloured blocks, even if, following Loyer (2015), those same blocks can be thought of as a kind of avatar *qua* 'enhanced cursor'; whereas it is possible for them to imagine, or at least imaginatively to reconstruct, the world and lore of Cyrodiil and to identify with both their character and the NPCs with whom they interact. There is,

[11] Hancock (2002) suggests that the cutscene serves 'to make a game's world more real – not just by telling a story, but also by reacting to the player, by showing him the effects of his actions upon that world and thus making both the world more real and his actions more important'.

[12] Tavinor makes a similar point, while discussing fictionalism in videogames, when he observes that 'videogame chess, Sudoku, and solitaire do not seem to present a fiction that one is playing these games in the sense that Oblivion presents a fiction that one is fighting a goblin or exploring an ancient ruin' (2012, p. 187).

then, a useful distinction to be made between, on the one hand, 'being immersed in' and, on the other hand, 'being in an immersive experience' (in the vernacular sense often used by players and game developers); albeit acknowledging that the latter is encompassed by the former.[13]

The first notion ('being immersed in') relates to the sense of being sufficiently focussed on a task—cognitively and/or motorically—such that one becomes less aware of one's surroundings, the passing of time, and external sensory stimuli. Seah and Cairns (2008) describe this as 'The sense of being lost in the game where players lose awareness of their surroundings and their day-to-day concerns' (see also Whalen 2004). This is 'preoccupation', and not so different to being immersed in a crossword or in any complex task requiring high levels of concentration and directed attention—and so would not appear to constitute a *defining* element of videogaming experience. This is not to suggest that such activities cannot activate powerful emotional responses (e.g. pleasure, a sense of achievement, frustration, anger etc.—one need only think of so-called 'controller rage'); but such responses are hardly unique to videogames.

Hence, the second notion ('being in an immersive experience') relates to the sense of identifying with, of somehow 'being in', of having 'agency' within, an imaginary or virtual world. There have been many attempts to categorise such experiences (see, for example, Brown and Cairns 2004; Calleja 2007, 2011). Adopting a dynamic/developmental, rather than taxonomic, approach, Ermi and Mäyrä (2005) proposed the well-known and oft-cited SCI model: a three-level model incorporating 'sensory immersion', 'challenge-based immersion' and 'imaginative immersion'. The end- or apex-stage has parallels with the notion of 'flow' associated with Csikszentmihalyi (2002), a concept adapted and utilised by Wiebe et al. (2014) in their attempt to develop a methodology for measuring engagement (see also Zhang and Fu 2015). More recently, Calleja et al. (2016) seek to discuss affective involvement by adapting the 'bottom-up experience triangle' from cognitive psychology: wakefulness > attention > involvement > incorporation.

In proposing 'that immersion requires concentration, a sense of challenge, control over the game and finally, emotional involvement and real-world dissociation', Jennett et al. (2008) seek to distinguish it from the concepts of 'flow', 'cognitive absorption', and 'presence'. They suggest that, while flow overlaps with immersion—in fact they posit immersion as a precursor for flow, rather than the other way around—it is an optimal, and consequently rare, experience. In addition, they conceive presence as primarily a psychological state-of-mind, as a sense 'of being' in a virtual world; as opposed to immersion, which they argue is an experience in time (see also Nacke and Lindley 2009). This aligns with Hartmann et al. (2015), who define (spatial) presence as 'the subjective experience of a user or onlooker to be physically located in a mediated space, although it is just an illusion' (p. 117). Here, the game(world) has become the primary ego reference frame. The fact that such a world will likely be fantastical misses the point. It is the coherence of the game-

[13]Calleja (2011) distinguishes between 'immersion as absorption' and 'immersion as transportation'.

world, as presented, and represented in the player's mind, which matters. This is why film-goers or videogame-players will often be critical of elements that appear to contravene the established 'lore' of a film or game (whether *Star Wars* or *Star Ocean: Integrity and Faithfulness*).

When considering the game-world, the notion of the 'magic circle' requires little summary. Salen and Zimmerman (2003), drawing on Huizinga (1955), proposed that games take place in a delimited, separate space—be that material or ideal—governed by the 'rules of the game'. The primary criticisms of this model—criticising the 'magic circle' became its own industry in the mid-2000s (see, for example, Lammes 2006; Taylor 2006)—are directed at its seeming failure to account for the permeability of the purported liminal boundary; and so its ignoring the reciprocal transfer of experience and affective motivation between 'game world' and 'real world' (see, for example, Calleja 2012; Stenros 2014). A sport (qua rules-based game) can be understood purely in terms of its rules; but this is not why millions of people play or watch football. For example, the *English Premier League* is marketed, presented, and broadcast as though it were an 'epic' tale, replete with personal stories, heroes and villains, 'plot twists', even myths etc.—it is about rather more than the 'rules of the game'.

Similarly, when I agonise over a moral decision in *Mass Effect 3* or determine to stop at red light in *GTA IV*, it is obvious that I am enacting ethical and behavioural prerogatives neither inherent in, nor explicitly given to me by, the 'rules of the game'; rather, I am transferring or projecting into the game attitudinal concerns from my 'real life' experience. Of course, it is more difficult to identify the mediating border-crossings in the magic circle's porous membrane when I play *Sonic the Hedgehog* (am I really driven by a moral imperative, or motivated by an emotional desire, to vanquish the evil Dr. Robotnik and prevent his collecting the six Chaos Emeralds that would allow him to turn all the animals of South Island into robots?). This perhaps explains, in part, the comments of Eskelinen and other ludologists in the late 1990s and early 2000s: side-scrolling videogames involving pixelated Italian plumbers or blue hedgehogs would seem to be more about goal-oriented ludic activity, wherein the goal is, in a sense, self-defining (reach the end of the level). However, in a game such as *The Last of Us*, the end of a level, or the game, is not a goal in and of itself, more a consequence. Joel and Ellie are not mere avatars ('cursor substitutes') like Mario; the gameplay is deeply embedded within a complex and emotionally-sophisticated story, and the final cutscene (which flows seamlessly from the preceding gameplay) is fundamental, as is the underscore.[14]

[14] This highlights one of the major challenges in videogame studies: the more abstract or reductive the definition, and the more it resembles a kind of 'Weberian ideal type', the less one can justifiably predicate of the remarkably diverse range of exemplars one thereby wishes to encompass.

Music and Immersion

What is notable about the range of theoretical models concerned with immersion in videogames is that very few of them pay much attention to music and sound.[15] This is despite the fact that the completeness and consistency of multisensory information obviously includes sound and music; and anyone who has played a videogame without its intended music or sound (where it is possible to effect this) will testify to a radically altered experience. One reason for this is that the elements of multidisciplinary engagement are not symmetrically accessible—it is much easier for a musicologist at least to engage with the conceptual models encountered in media studies, game studies, or cultural theory etc., than it is for a non-musicologist to acquire the technical musical literacy required to engage with music in meaningful analytical detail.

Hiding in Plain Hearing

An additional reason, in terms of explaining its omission from general models, is captured by Berndt and Hartmann (2008), when they note that music's 'tremendous importance only becomes apparent by muting it and experiencing the disillusioning lack of this unconscious something' (p. 126); similarly, Lankoski (2012) observes that 'One needs only to turn off the sounds from the *Silent Hill 3* or *Thief Deadly Shadows* to see how the games lose their emotional impact. Music and beauty are similar to empathic engagement, because they cannot be (fully) explained by the goal-related emotion theories' (p. 40). Like music in film, it works precisely by virtue of the viewer/player not being aware of how it is working.

In part, (non-diegetic) music is accepted, even expected, because videogame players' expectations are often informed, inter-medially, by filmic convention (the underscore to *Red Dead Redemption* does not comprise the kind of 'authentic' music that would actually have been heard on the American frontier in 1911, but rather the kind of scoring associated with Morricone's music in Sergio Leone's Westerns of the 1960s). Another way of putting this is that, while empirical and psychological studies (which have tended to dominate attempts to measure 'flow' or 'immersion') afford useful data, they need to be accompanied by, and combined with, both musicological and semiotic analysis.

[15] For example, Calleja's (2011) book-length study of immersion and incorporation does not actually mention music (other than in a couple of analogies).

A.L.I.

One of the more developed attempts to relate music specifically to immersion is provided by Van Elferen (2016), who, citing the relatively sparse and otherwise vague reference to music in theoretical models of immersion (as noted in the preceding section), proposes what she terms the ALI model—(musical) Affect, Literacy and Interaction.

Affect is defined as 'the personal involvement in a given situation through memory, emotion and identification' (p. 34); and the role (or power) of music in this respect is obvious, even if challenging to quantify (though see some of the other contributions to this volume). Similarly, Zehnder and Lipscomb (2006) suggest that music is used to 'enhance a sense of immersion' by 'cueing narrative or plot changes, acting as an emotional signifier, enhancing the sense of aesthetic continuity and cultivating the thematic unity of a video game'.

Interaction relates to 'interaction with and through music', whether in actual 'music games' (such as *Rock Band*) or in the more general sense of dynamic scoring. I would part company with Van Elferen at this point, since her definition is too literal. It is a category error to place, under the single umbrella of 'musical interaction', such clearly different, and functionally different, instances as music and rhythm-action games, in-game player-based track selection (*GTA*, *Fallout*), and dynamic composition (which itself can refer to any of: the segue/fade from one track to another—horizontal sequencing—in response to a trigger-point; real-time dynamic change—vertical layering—within the 'same' track; or even procedural generation).

Literacy refers to 'habituated practices' of interpretation and is often inter-textual and trans-medial (whereby a videogame score may utilise well-known tropes from other media, such as film). Not all of Van Elferen's examples are apposite, however. She suggests that RPGs are typically scored in an 'epic' style comparable to epic and fantasy films. There are sufficient examples of this not being to case to query the assertion. For example, other than their title themes—which do align with the musical semiotics of an epic-fantasy film-score—the overwhelming bulk of the in-game music in *The Elder Scrolls* series (such as *Oblivion* or *Skyrim*) is delicate, ambient, and low in the mix; rarely, if ever, do we encounter the type of soaring score that accompanies the sweeping high-angle shots in, say, *Lord of the Rings* (the film trilogy). The *Souls* franchise reserves music only for major 'boss battles'. Similarly, *Dragon's Dogma* includes extended periods without underscore, typically utilising sporadic non-combat music as a signifier for specific locations; the same is true of the *Sacred Angel* series; and *The Witcher* franchise tends to employ gentle and unobtrusive string-based folk-like music with Slavic overtones. In effect, these open-world 'Westernised' fantasy RPGs have developed their own idiom. Conversely, where one encounters more persistent (pseudo-)epic music in RPGs—for example in *Final Fantasy XIII*—it tends to adopt a style that is instantly recognisable from JRPGs (melodramatic, overblown, and combining an eclectic mix of Romantic orchestration, jazz, and techno-rock).

What is therefore missing, perhaps, from Van Elferen's model is an awareness of 'functional context' (or 'game-style context'), especially in respect of immersion. The pared-down underscoring found in the 'Westernised' RPGs listed above is due to their fully open-world, exploratory setting, since anything overtly epic or clearly looping would draw too much attention to itself and actually damage immersion (even the combat music tends to be understated by videogame standards); at the same time, each carries subtle inflections that speak to the respective naturalistically-realised (if fantasy) setting—ethereal and magical in *The Elder Scrolls*, East European Medieval folk in *The Witcher* etc. Conversely, *Final Fantasy XIII* is significantly more linear and combat-oriented, with elements of its stylised presentation drawn from anime and steampunk; its exploration episodes (which remain directed and so linear in effect) are sometimes accompanied by a gentler style, but even here are more obviously shaped (melodically and harmonically) and often carry leitmotivic references to other musical themes encountered in the game.

Music and Cutscenes

Van Elferen's model (whether modified or not) could plausibly incorporate cutscenes, but the focus is evidently on 'gameplay'. This is because, where cutscenes are concerned, a common perception is that they are 'devoted solely to advancing the narrative', and so can be scored 'just as a film is scored' (Phillips 2014, p. 82). This is perhaps too simple—on both counts. Certainly, cutscenes proffer an efficient and information-rich aid to narrative-progression, and this is often their purpose (and, again, this also has ramification for VR). Yet, as noted above (see also Ip 2011a, b), they can clearly do more than this: develop character, promote emotional investment, contextualise choices to be made by the player, reveal the consequences of choices already made, and even add to, rather than detract from, immersion.[16]

Musical Dramaturgy

In respect of the functions just listed, one might point to extended music-centric cutscenes such as the renditions of 'The Dawn Will Come' in *Dragon Age Inquisition* or 'The Wolven Storm' in *The Witcher 3*—both of which incorporate originally-composed yet diegetic music (one of the less common forms of videogame music).[17]

[16] It should be noted that cutscenes would historically—at least from the PS1 era onwards—also serve a technical function, by allowing the game to download the next 'game area', internally, in the background, while the pre-rendered sequence was simultaneously streamed to the screen. In that sense, cutscenes could also aid immersion, at least in relative terms, by removing the need for, or at least shortening, static 'load screens'.

[17] As one might expect, the most common form of videogame music is originally-composed non-

These are not obviously concerned with narrative *advancement*. Rather, they are moments of repose and reflection—if anything, they are caesuras in the narrative progression, a kind of musically-realised narrative consolidation. In addition, they contribute significantly to world-building and setting. 'The Dawn Will Come', modal (d-dorian) and with a sacred Medieval resonance, is set for collective voice, though is initiated by a single NPC character (a religious figure) whom the others gradually join in singing—a classic epiphanic trope (as seen for example in the film *Twelve Years a Slave*). In contrast, 'The Wolven Storm' is for solo voice and lute, delivered to an audience in a tavern, and has a secular folk resonance (wavering between modal and diatonic presentation), in the manner of Troubadour repertoire—it is also replete with intra-textual lyrical references ('berries tart, lilac sweet' recalling the 'lilac and gooseberries' of the tutorial introduction; 'The wolf I will follow' clearly referencing the player-avatar, Geralt, who belongs to 'the Witcher school of the wolf'). Consequently, not only do these scenes perform a similar 'caesura' function in both games, but their music reflects and reinforces the respective setting, tenor, and context of each. They are not (functionally) dissimilar to the striking and remarkably affecting appearance of Jose Gonzales' 'Far Away', as non-diegetic sound-tracking, in *Red Dead Redemption*, at the point where Marston, the player-character, rides alongside the Rio Grande in a sequence which effectively bifurcates the principal 'acts' of the game.

Similarly, the cutscene around the death of Aerith in *Final Fantasy VII* extends temporally far beyond what is required for narrative development: the plot points—Aerith is dead, Cloud is distraught and seeks revenge—could have been conveyed in a cutscene lasting a tenth of the time. Its primary purpose, one in which the music plays the central role, is concerned with emotional affect and investment (many players report being moved to tears). Rather than *only* advancing the narrative, it serves to reframe and retrospectively recontextualise what has happened up until this point; and this is achieved *musically*, via the poignantly pared-down (Midi-realised) piano version of what the player has come to recognise as Aerith's theme.

The theme articulates a duality on multiple levels. The first four bars gently undulate between a muted I and (minor) v - muted insofar as I is clouded by the suspended second, and v is realised in second inversion. One might observe that the suspended E is the same similarly suspended note which, as discussed below, opens the game's first cinematic sequence (as the camera cuts to Aerith), just as the major-minor duality mirrors the larger-scale E-major/a-minor harmonic trajectory established in that same opening sequence. The consequent phrase is marginally more dynamic, with inner-voice motion and the transfer of v to root position. The ascending motif (against I) is answered by the descending motion (against v). Aerith is

diegetic underscore; next is pre-existing music employed in a diegetic context (as in *GTA* or *Fallout*); originally-composed diegetic tracks, usually songs, are less common, though typically found in RPGs; the least common type, unsurprisingly, is pre-existing music employed in a non-diegetic function (since its intertextual semiotic signification too obviously draws attention to itself and so compromises immersion, with instances often limited to moments of obvious and intentional parody, such as those found in *Far Cry 3* or *Saints Row*).

both the symbol of hope (and purity) and the victim of a seemingly arbitrary fate—a duality reflected by the piquant alternation between ascending major and descending minor.

The opening period is extended via a detour through the flattened submediant and a modally-inflected minor subdominant—in effect a chromatic, pedal-like, under-carriage to the sustained tonic root in the melody, which serves both to retain the major-minor alternation and also to provide a sense of non-dominant cadential momentum and repose. The counter-subject (second period, beginning with the anacrusis in b.10) is realised via a plaintive, and literally pathetic, 'oboe'. As in much of Uematsu's music—itself an eclectic mix of late nineteenth century classical-romantic practice, jazz-derived harmonic colouring, and modal transformation—clear dominant-tonic function is mostly eschewed throughout; and the plagal implication of the various cadential moments (or half-closes) serves to reinforce the spiritual implication of Aerith's role and her fateful (and fated) sacrifice.

What is especially striking about the use of this theme during the 'death scene' is that it continues into, throughout, and beyond, the subsequent battle scene (with Jenova LIFE—a particular iteration of the repeatedly encountered Jenova 'boss' enemy). In one of very few battles to omit the standard RPG post-win 'victory pose', the fight with the poignantly-named 'LIFE' version of Jenova is imbued with the fact of Aerith's death, by virtue of the uninterrupted continuation of her musical theme. Rather than a dramatic high-point, the boss-battle itself is rendered contextually secondary—by the music.[18]

The key point is that, even if one insists on a clear—I would propose implausible—binary distinction between 'gameplay' and 'cutscene', it is very often the music which performs the 'suturing' or 'integrative' function. Common to the majority of audio-visual theorisation—in respect of film or videogames—is the notion of an 'emergent' quality or property, deriving from what Collins (2013), borrowing from Chion, refers to as 'synchresis'. Accordingly, 'The theory of multisensory integration holds that there is a synthesis or binding of information that occurs between modalities in which the information that emerges could not have been obtained from each modality on its own' (p. 27). This explains why almost all cutscenes—certainly those of emotional-dramaturgical import—incorporate music.

[18] Summers observes, in relation to Halo: 'One way in which this link [between first-person action and the broader plot event] is achieved is by challenging the strict separation of interactive (first-person) gameplay and non-interactive (third-person, plot-expository) cutscenes. Musical material that is heard in the cutscenes is reprised during the central gameplay, connecting the two domains, and cues that transgress the game mode boundary serve to smooth over the change of interactive mode'. (2012, p. 145)

Congruence

Most scholarship on the role of music in videogames tends to presuppose some form of dramatic, kinetic or synesthetic congruence. This is similar to the approach adopted in much film-music scholarship. For example, in discussing *Jaws*, Wingstedt et al. (2001) note how the musical expression is congruent with the overall narrative. The almost naturalistic main theme 'heightens immersion as well as modality' (p. 202). Similarly, Lerner's (2014) analysis of music and sound in *Super Mario Bros* is predicated on demonstrating audio-visual and audio-ludic congruence. Gasselseder (2014) suggests that videogames generate immersive qualities by 'keeping structural and expressive features of stimuli congruent while playing' (p. 18); and that 'If the multisensory stimuli indeed match congruently to the hypotheses of perception, an intensified allocation of attentional resources to the media content arises' (p. 19). Likewise, Williams et al. (2016) observe that 'Combining emotionally congruent sound-tracking with game narrative has the potential to create significantly stronger affective responses than either stimulus alone – the power of multimodal stimuli on affective response has been shown both anecdotally and scientifically'.

Most videogame cutscenes do indeed adopt what might be described as an orthodox approach to the conjoining of congruent audio/musical and visual elements. A good example is afforded by the aforementioned cinematic sequence which opens *Final Fantasy VII*. After a segment in which the 'camera' spirals across a starry sky, the sequence cuts to a figure (a 'flower girl', whom we later learn to be Aerith). The single high 'string'—a classic signifier of tension/anticipation—is held until, in something akin to 'Mickey Mousing', the first melodic material, an unsettling chromatic motif, accompanies the character's turning her head, seemingly 'to camera'.[19]

The high 'string' is sustained, while additional synthesised samples gradually enrich the texture, as the visual sequence follows the character out onto the street; at which point the 'camera' pans away and upwards to afford a birds-eye view of the game's futuristic steam-punk setting. Then, as the game legend appears on screen, cued by a 'timpani roll', the principal thematic material sounds in the 'brass'—a muted, yet portentous and clearly diatonic phrase, which is set against strings and a quietly percussive bass, all now coalesced into a clear and unambiguous E-major. The phrase repeats, more emphatically, at the octave and with a brighter timbre, as the harmony moves sequentially to VI.[20] This represents the musical and figurative climax, intimating a sense of forthright and implicitly heroic purpose.

---

[19] References to instruments are placed in 'inverted commas' to reflect the fact that these are MIDI-realised sounds

[20] The obvious allusion to the Vangelis track, which accompanies the opening sequence in the film *Blade Runner*, is unlikely to have been coincidental, especially if one considers not only the similarity in art direction—a slow camera pan over a futuristic city-scape at night, accompanied by motivic brass fanfares set against sustained high strings—but also several narrative and thematic parallels.

However, no sooner is the peak reached—both musically and visually (the camera at its 'highest' long-distance angle)—than for it to be immediately countered by an audio-visual descent that is both literal and metaphorical. As the camera pans down and, cut against interspersed shots of an approaching train, zooms-in on the cutscene's final shot, the music abruptly shifts to a-minor and, via a concatenated stretto of arpeggiated descents, remains there. In a relatively simple and efficient manner, the music accompanying the opening sequence has established, via congruence, key narrative and emotional tropes: anticipation; hope; foreboding.

Having noted this, and while it is true that emotionally congruent sound-tracking can produce a more powerful response than either element alone, it is also the case that (even more) powerful affective responses can be generated by specific kinds of *incongruent* juxtaposition.

Incongruence

Film music scholarship has long recognised the concept of audio-visual 'counterpoint': those instances where the symbolic, emotional or dramatic signification of the music appears not to align with—is *incongruent* with—the superficial signification of the visual sequence with which it is conjoined. For example, Bordwell and Thompson refer to 'ironic contrasting' (2008, p. 302); Chion describes such music as having an 'anempathetic' relationship to the visual (1994, pp. 8–9); and, similarly, Smith (1999) distinguishes between what he calls 'affective congruence' and 'polarisation'.[21]

Filmic examples of audio-visual incongruence are well-known. They include, for example, many of the scenes in *Clockwork Orange* or the infamous 'torture scene' in *Reservoir Dogs*. Similar techniques are often encountered in the work of documentary film-makers, usually serving to communicate an ironising moral-political point that inheres in neither image nor music alone, but rather is an emergent property of their combination (for example, 'What a Wonderful World' in Moore's *Bowling for Columbine*). They are also common in (anti-) war-films, from 'We'll Meet Again' in *Dr. Strangelove* to Barber's 'Adagio for Strings' in *Platoon*. It is necessary to observe, as do Willemsen and Kiss (2013), that 'irony', even when broadly conceived, does not adequately describe the artistic intention behind, or the consequent effect of, all instances of audio-visual incongruence. One striking, and decidedly non-ironising, example occurs towards the end of the battle-scene which

[21] The device has also been interpreted with ends other than straightforward theorisation in mind. For Adorno and Eisler, audio-visual counterpoint represented a normative aesthetic, whose purpose was to undermine or deconstruct the superficially seamless, and ideologically suspect, presentation of a unified and falsely affirmative artistic whole (or Gesamtkunstwerk). For others, the same device serves to re-frame that which is presented visually, by opening-up a space which allows for humour, irony, or pathos; or, as in the theoretical framework presented by Lissa, creates a 'dialectical unity', an audio-visual 'thesis-antithesis-synthesis' counterpart to Eisenstein's montage theory.

opens Scott's *Gladiator*, in which a slow, elegiac string version of the marshal main theme is juxtaposed against the brutal and violent climax to the battle.

However, Berndt and Hartman suggest that 'the only established functions [of music] in games are those providing the feeling of immersion into the scenario and a superficial dramatization of action scenes. … Hence, interactive media are far behind their possibilities and even behind ancestors. Especially the contrapuntal functions seem completely unknown to game developers' (2008, p. 130). This is not true.

A good example appears in *Gears of War 3*. At the end of Act 3, Dom, a major character in the series, sacrifices himself in order to save his team. The cutscene depicting the action—as Dom drives his explosives-laden vehicle into the enemies and the resulting explosions and carnage are depicted in pseudo slow-motion—is accompanied by a soft instrumental version (piano, then with the addition of synthesised string) of 'Mad World'—originally by Tears for Fears, but here an instrumental version of the Gary Jules and Michael Andrews' cover version, which also accompanied the final montage in the film *Donnie Darko* (which the game's developers cited as among their favourites). Of course, the intensity of inter-textual resonance will depend on the awareness of the player: some will recognise the track as referencing the conclusion to *Donnie Darko*; some will recognise the song as 'Mad World'—rendering a poignant lyrical commentary on the conflict inherent to this war-based videogame; others will make neither connection. However, even in the case of the latter, the incongruent juxtaposition remains effective and affecting.

A similar technique is employed in the middle section of the concluding cutscene sequence that closes *Mass Effect 3*. The player having made their final decision—there are up to three choices—the game closes with an extended montage.[22] As Shepard (the player protagonist) literally makes their final leap, the sequence cuts between sepia-tinged footage of companion NPCs, ongoing fighting, and the final destruction of the galactic relay system. As this climactic and dramatic sequence unfolds, it is, however, accompanied by the soft, slow-tempo, pianistic 'An End, Once and For All' (an original composition for the game, by Clint Mansell). The piece is clearly located in the sound-world of Ludovico Einaudi: non-functional harmony (here centred around an e-dorian scale-set); triadic identity dampened by the addition of seconds and sixths (a single chromatic inflection appearing twice in the opening 20 bars); and the repetition and developing variation of sparse arpeggiated motives. The (formally incongruent) music here serves three functions: firstly, it conjoins visual 'flashbacks' with musical 'flashbacks' (the ascending-descending fifth motif, which makes its first appearance in b.13); secondly, it provides an aural continuity, a musical pedal, suturing the cross-cutting visual shots

[22] The ending to *Mass Effect 3* was a source of significant contention, even anger, among gamers—especially those who had invested in the entire series. Many were dissatisfied with the three choices available, with the lack of a 'perfect ending' possibility, with the seeming inconsequence of the many choices they had made throughout the game, and with the fact that the three cut-scenes—each reflecting the final decision—appeared almost identical. Such was the outcry that the developers released an enhanced and expanded final game sequence in the form of free DLC (downloadable content).

and lending a summative feel to the temporally and physically disjunctive visual cues[23]; thirdly, and most importantly, it serves to retrospectively re-frame the game just completed.

The Final Cut

Players not only (re)construct a mental map of the physical world represented in (VR and non-VR) virtual space, but also a chronological and temporal frame of reference—whether that be the complex and lore-based historical backstory to *White Knight Chronicles*, or the fact that Stage 2 logically follows Stage 1 in *Ghost and Goblins* or *Revenge of Shinobi* (both of which use pre-stage scrolling maps to lock together a sense of geospatial and chronological progression).[24] It is the macro-level abstracted/inferred timeline which allows players to make sense of the flash-backs and chronologically non-sequential episodes in a game such as *Uncharted 4* (just as they would when watching *The Walking Dead* or almost any film by Quentin Tarantino). Cutscenes are therefore central both to creating the sense of a 'before and after' context, and of progression, for the events played-out in the game itself; and also to anchoring temporal disjunction in a non-sequential videogame narrative (or the parsing of 'discourse' and 'story' as literary theory would put it). In early (1980s) games, selecting 'start' on the title screen would typically see the player simply begin *in medias res* (although some videogames, such as *R-Type*, might include a short perfunctory 'launch sequence'). Almost all later and contemporary games tend to incorporate some form of 'pre-framing' contextualisation—usually via cutscenes. O'Grady goes further, proposing that '… cutscenes may in fact warrant reconsideration for their ability … to reconcile temporal disunities that would otherwise overthrow a player's sense of interactive control and, ultimately, her sense of "being" in the gameworld' (2013, p. 104).

In the overwhelming majority of videogames, the 'actual gameplay' represents an indeterminate means of moving from one pre-determined point to the next—in fact, 'games' that lack any purpose or goal are generally not considered games, but rather 'experiences'. Whatever strategy I adopt in killing the NPC enemies in a given mission/level in *Dragon Age: Inquisition*, the narrative will 'wait for me' (and remain framed by the *same* cutscenes)—even if I die and must restart the level. As Thon suggests: 'Contemporary video games commonly use a variety of strategies of prototypically narrative representation such as cut-scenes or scripted sequences of events, and the events thus presented are generally *highly determined* before the

[23] Something similar occurs in the library scene in Finch's film *Se7en*, where Bach's 'Air on a G-string' plays in its entirety—moving from diegetic to non-diegetic in the mix—as an extended montage sequence, without dialogue, cuts between shots of the two lead protagonists in different locations and at different times.

[24] Many contemporary/app-based casual 'puzzle games', such as *Candy Crush* or *Cut the Rope*, still retain the conceit of progressing along a 'map' route.

game is played' (2016; my emphasis). Just as the music accompanying the 'game-play' will be cued to the mechanic in action (looping, dynamic etc), so it is the music within the framing cutscenes which carries, invariantly, the narrative and emotional 'punch'—audibly marking-out the dramaturgical rungs on the ladder of teleological progression. By whatever means the player gets there, *that* Aerith cutscene will always play, *that* Dom cutscene will always play. A fallacy in some videogame scholarship is an exaggerated emphasis on the 'indeterminacy through interactivity' of player intervention. The apparent ludic freedom of immediate 'moment to moment' player-determination is almost always subordinate to the constraints of the higher-level, and implicitly teleological, structuring (whether that be reaching the end of the level in *Gradius* or devoting an hour to 'grinding' experience points in *World of Warcraft* in order to defeat a given enemy).

An Extended (Virtual) Coda

As has been shown, audio-visual congruence, and especially incongruence, are employed in some of the most emotionally affecting and memorable cutscenes in videogame history. However, such instances—often at the end of a videogame or on the death of a major character—almost exclusively utilise pre-rendered cutscenes, in which the player has no direct control (not even camera or view-point control), since the framing and dramaturgy require cinematic direction for maximum effect. But in VR one cannot physically prevent a player from turning their head (since locking the view, irrespective of whether or not they do turn their head, risks shattering the illusion, not to mention potentially inducing motion sickness). The issue for VR is how to replicate the purpose and affective intention of such cutscenes; and the risk, in the context of VR, is the 'wrong' kind of incongruence. Hence, two of the primary challenges in VR—in respect of traditional models of both composition and story-telling—relate precisely to the three elements conjoined in the preceding discussion: the role of cutscenes (in the sense of their conventional presentation and function in non-VR games); the role of music (in the non-diegetic sound-tracking sense); and how each of these is bound-up with immersion.[25]

[25] Given that VR is a recent commercial reality, VR games are in their infancy, the technology and techniques are still in an early stage of development, and there is—obviously—little extant academic scholarship dealing with VR videogames, so this section is necessarily more provisional and positional in its observation.

Cutscenes in VR

Whether or not one considers cutscenes to be immersion-breaking in general, it would appear that VR exacerbates their potential to be so. One might therefore expect a greater emphasis on incorporating narrative, backstory, and lore within the 'real time' in-game context. Historically, a number of conventional games have adopted this approach, such as *Bioshock.* While Madigan (2010) argues that 'Bioshock's audio logs kind of hurt the game's otherwise substantial immersion … who the heck records an audio diary, splits it into 20 s chunks, puts them on their own dedicated tape players and then wedges those tape players into various corners of a public place?', this tends to overlook the fact that all videogames incorporate similar conceits. In fact, the 'collect and piece-together' trope is a staple of innumerable videogame genres; and these need not imperil immersion (any more than *Call of Duty* is less immersive because we know that in 'real life' a soldier would not expect to find regular supplies of ammunition conveniently placed along their path or expect to cure multiple gunshot wounds by briefly crouching behind a wooden crate).

A similar approach will likely see 'cutscenes'—e.g. conversation, dialogue, other contextualising events—play-out from the perspective of the player. At the same time, this could lead to the VR equivalent of the *Half-Life 2* conundrum. The latter featured very few cinematic pre-rendered sequences or even enhanced 'real-time' assets. Rather, almost every narrative or expositional scene was played-out in the player's game-space. It not only allowed camera control, but in most cases also allowed the player to move around; and is often cited (and lauded) by players for this reason. However, it came at a certain cost. The mechanic naturally required that the player be confined to a small space (usually a locked room—a kind visible 'invisible wall'); the player could turn the camera away from the NPCs engaging in activity or dialogue, and could perform actions such as jumping on furniture or even 'shooting' (to no effect) in the direction of the NPCs (who did not react). Yet, even if this were immersion-breaking, it was the player who determined to do this. Hence, in general, and in a virtuous circle of self-reinforcement, the integrated sequences and retention of character control created a more seamless and, for some, more immersive experience, which itself tended to encourage the player to observe the scene as presumably intended. In VR, spatialised audio will play a crucial role in drawing the player's visual attention in the 'right' direction.

One can assume that 'cutscenes' in VR will adopt similar techniques. This is apparent, for example, in *Batman Arkham VR*—one of few VR games, so far released (at time of writing), to attempt a naturalistically-conceived, first-person, narrative-driven experience. In fact, *Arkham VR* tends to blur the presentational boundary between 'cutscene' and gameplay. The game takes place in a series of fixed settings (the player effectively 'rooted to the spot', but with freedom of viewpoint control and mechanical interaction). These include, for example, the opening scene, in which the player-protagonist witnesses (in keeping with the lore) the murder of his parents; or the following (quasi-tutorial) scene in which the player converses with Alfred, the butler. This mechanic serves partly to evade what Warnke (2015) sees as the fundamental paradox of VR: 'The striking contradiction between the freedom of movement it promises and the physical 'dispositif' of its users seems to be at its phantasmic core. While cybernauts effortlessly travel through an endless virtual world, they are in a kind of iron maiden that encloses them on the spot. Immobile freedom, or local boundlessness—this seems to be the paradox of virtual reality as a technology of immersion'. Another way of putting this is that VR achieves a new type of immersion at the same time as potentially affording new ways of undermining it (such as the often-criticised 'teleportation' method of locomotion typically employed in early VR releases).

It would seem likely, despite the semantic discussion which opened this chapter, that we will need an expanded vocabulary with which to describe and theorise the way in which developing VR mechanics accomplish the work otherwise associated with conventional cutscenes in non-VR videogames.

Music in VR

It is well established, as discussed above, that both film-watchers and game-players not only typically expect non-diegetic underscore—such that films or games without it tend to stand-out for that reason—but that it normally enhances emotional engagement and immersion. At first glance, this appears paradoxical. In the 'Pigs in Space' skits in *The Muppet Show*, the muppets would often look up and around in confusion as the continuity announcer introduced the episode: an example of 'breaking the fourth wall'. In theory, as we wander across the wilds of *Skyrim*, we would not expect to hear a gentle ambient score 'from nowhere', which then segues dynamically into combat-music as we move to within the game-defined proximity-area internally programmed to trigger it.

However, rather than immersion-breaking, the music arguably functions in the opposite way. It is only when the algorithmic intention breaks-down that we notice it: e.g. when a proximity-trigger fails to take-account of a wall or vertical drop, and so triggers the 'enemy-near' combat theme, even though there is no possibility of engaging; or when a player acts 'unexpectedly', such as repeating moving back and forth rapidly between audio trigger-points, leading to an ungainly cutting back and forth between musical themes intended to smoothly segue or cross-fade from one to the other.[26]

This is not to say that some games do not derive their apparent immersive effect from the converse direction. Horror and survival games will typically opt for an

[26] Of course, some games, such as *Red Dead Redemption*, seek to avoid this by utilising a dynamic form of underscore based on vertical layering. This can react for more quickly, and subtly, to player-action and game-state, albeit thereby restricting some of the parameters (modulation, melodic development, and tempo-change) central to character-representation and emotional narrative.

under-score in which the distinction between ambient or environmental sound and conventional musical presentation is blurred, often in favour of the former: anything from *Dead Space* to *Limbo* are good examples of this. Similarly, the football simulation *FIFA* restricts original music to its menus, whereas 'in-game' audio is limited to the pre-existing tracks one normally hears as the teams walk-out, faux TV presentation effects, and increasingly realistic and 'authentic' (if clean) versions of football chants.

However, the dominance of non-diegetic sound-tracking in both film and also the majority of videogames is in part due to the fact that almost all of us have become acculturated, from the earliest age, to expect it in a presentation on a screen ('3D games', as many now are, are still presented on a 2D screen some distance away). Arguably, the music closes that distance, whether by drawing-us into an open-world or action-adventure game; or by helping to close-out the external environment that might otherwise distract us. Similarly, as Kamp notes: '... background music is phenomenologically nondiegetic. In this nondiegetic role, music can function like the "phenomenal filter" ... altering the player's experience of objects they attend to while remaining "invisible"' (2013, p. 240) (see also Moffat and Kiegler 2008).

In the case of VR, the question, or challenge, is whether the inclusion of non-diegetic underscore is more likely to provoke a 'Pigs in Space' moment. It is already apparent that musical underscore is either absent or less obvious in the first tranche of VR game releases—albeit this may simply be due to the nascent state of the technology and the fact that a reasonable number of early VR 'games' are more akin to tech demos. Nevertheless, as Phillips observed in an interview:

> For music in virtual reality, the big discussion right now is whether music should be localised to the physical world, or whether it should remain as more the background accompaniment that follows the player wherever they go. The issue is the sense of presence that virtual reality gives the player, as they are more conscious of a sense of personal occupation, since they are inside the world, exploring. A problem we grapple with and worry about is the idea of music hovering over their shoulder and following them, which starts to feel like the 'Music God'. (2017)

To some extent, this may be mitigated by the fact, as already noted, that we are primed to expect, or accept, non-diegetic music in audio-visual presentation. This explains why the aforementioned 'Far Away' in *Red Dead Redemption*—which, as a non-diegetic pre-existing song, should be immersion-shattering—has the opposite effect (and is often cited in 'best uses of music in videogames' lists). Players are attuned to adopting different modalities of engagement and to suspending (various types of) disbelief. However, there are likely to be two tendencies, at least so in more 'naturalistic' VR games.

The spatial dimension becomes especially relevant to sound effects and environmental noise (point of sound-source relative to the player's head position)—at least if an authentic experience is to be created.[27] Many modern conventional games

[27] This is obviously not the same as speaker-based surround-sound or virtual surround-sound (VSS) headphones, since these assume that the listener is looking in a fixed direction i.e. at a screen. For this reason, '3D headphones', compatible with VR use, require a wired connection, since the rel-

already utilise dynamic 'real time' processing in order to create more acoustically realistic effects. For example, *Battlefield 4* employs a complex algorithmic patch simply to render footsteps relative to variables such as speed or surface (see Stevens and Raybould 2015). Similarly, in the scene near the beginning of *Bioshock Infinite*, where the player character approaches a carnival/funfair, a variety of (simple) techniques—proximity-dependent volume attenuation, reverb, and low-pass filters—are used to manipulate the sound of the collective singing coming from 'up ahead'. However, these examples assume the player is looking straight-ahead at a fixed screen (if I turn my head while wearing VSS headphones, the sound that was coming from 'the left' still sounds as though it is coming from 'the left'). Hence, this is highly relevant for first-person VR games in which diegetic audio-effect is central both to player orientation and also to drawing/directing their visual attention.[28]

Secondly, there will be an increase in diegetic source (as in *BioShock, Fallout,* or *GTA*); and/or a move towards a blurring of sound-tracking, sound-effect and circumambient sound (as in *Silent Hill* or *Dead Space*). The interplay of point-source 3D/binaural audio and pervasive/non-spatialised circumambient audio will be crucial. In fact, this is already apparent in a number of early VR releases. For example, *Theseus VR* very much follows the *Dead Space* template: an oppressive exaggerated/reverb-saturated ambient sound; high dissonant string stabs; and low (Zimmer-esque) bass/brass swells—its more percussive moments (realised in the manner of high-modernist experimentation) are generally reserved for combat. That said, *Theseus VR* is third-person; and the distancing effect therefore allows a relationship between player and controlled-character more in keeping with conventional videogames (a kind of *Gods of War* meets *Ico* in a virtual reality space). One early first-person release to include non-diegetic underscore is *Arizona Sunshine VR*. Effectively a zombie survival game, there is no persistent music, but rather, at various points, a stylised ambient-type accompaniment fades in (similar to an understated version of *Red Dead Redemption*). Meanwhile, *London Heist VR* switches between no music and full-blown tracks (reminiscent of *James Bond* films) during action/combat sequences. In short, it is false to assume that non-diegetic underscore is inappropriate or immersion-breaking *per se*—the player is still aware that they are 'in a game' and is still taking inter-medial cues not only from film but also from non-VR videogames.

VR does not obviously change the issues discussed prior to this 'coda'. Rather, it requires (and will require) alternative game-mechanics and 'conceits' to achieve the same. Just as the assumed greater 'immersion' of VR is potentially jeopardised by the constraints it must overcome, so the presumed lesser relevance of music and sound is confounded by the significant, if different, role it will come to play.

evant processing requires positional and movement data from the headset.

[28]Techniques and middle-ware applications, for generating real-time dynamic full 360o audio, are still in various stages of experimental development (the three major commercial VR systems each utilise proprietary technology).

References

Berndt A., Hartmann K.: The Functions of Music in Interactive Media. In: Spierling, U., Szilas, N. (eds.) Interactive Storytelling. ICIDS 2008. Lecture Notes in Computer Science, vol 5334. Springer, Berlin, Heidelberg (2008)

Bordwell, D., Thompson, K.: Film Art: An Introduction. McGraw-Hill, Boston (2008)

Brown, E., Cairns, P.: A grounded investigation of immersion in games. In: (Proceedings) ACM Conference on Human Factors in Computing Systems, CHI (2004)

Calleja, G.: In-Game: From Immersion to Incorporation. MIT Press, Cambridge, MA (2011)

Calleja, G.: Erasing the magic circle. In: Sageng, J.R., Fossheim, H., Larsen, T.M. (eds.) The Philosophy of Computer Games. Springer, London (2012)

Calleja, G., Herrewijn, L., Poels, K.: Affective involvement in digital games. In: Karpouzis, K., Yannakakis, G. (eds.) Emotion in Games. Socio-Affective Computing, vol 4. Springer, Cham (2016)

Cheng, P.: Waiting for something to happen: narratives, interactivity and agency and the video game cut-scene. In: Proceedings of the 2007 DiGRA International Conference: Situated Play (2007)

Chion, M.: Audio-Vision: Sound on Screens. Columbia University Press, New York (1994)

Collins, K.: In the loop: creativity and constraint in 8-bit video game audio. twentieth-century music **4**(2), 209-227 (2007)

Collins, K.: Playing with Sound: A Theory of Interacting with Sound and Music in Video Games. MIT Press, Cambridge, MA (2013)

Csikszentmihalyi, M.: Flow: The Psychology of Happiness. Rider, London (2002)

Donnelly, K.J., Gibbons, W., Lerner, N. (eds.): Music in Video Games. Routledge, London (2014)

Ermi, L., Mayra, F.: Fundamental components of the gameplay experience: analysing immersion. In: Proceedings of the 2005 DiGRA International Conference: Changing Views: Worlds in Play (2005)

Eskelinen, M.: The gaming situation. Game Stud. **1**(1). http://www.gamestudies.org/0101/eskelinen/ (2001)

Fahlenbrach, K.: Affective spaces and audiovisual metaphors in video games. In: Perron, B., Schröter, F. (eds.) Video Games and the Mind: Essays on Cognition, Affect and Emotion. Macfarland, Jefferson, NC (2016)

Fritsch, M.: History of video game music. In: Moormann, P. (ed.) Music and Game. Musik und Medien. Springer VS, Wiesbaden (2013)

Gasselseder, H-P.: Re-scoring the game's score: dynamic music, personality and immersion in the ludonarrative. IADIS International Journal on WWW/Internet 12(1), 17–34 (2014)

Hancock, H.: Better game design through cutscenes. Gamasutra. https://www.gamasutra.com/view/feature/131410/better_game_design_through_.php?page=2 (2002)

Hartmann, T., Wirth, W., Vorderer, P., et al. Spatial Presence Theory: State of the Art and Challenges Ahead. In: Lombard, M., Biocca, F., Freeman, J., et al (eds.) Immersed in Media. Springer, Cham (2015)

Howells, S.: Watching a game, playing a movie: when media collide. In: King, G., Krzywinska, T. (eds.) ScreenPlay: Cinema/Videogames/Interfaces. Wallflower Press, New York (2002)

Huizinga, J.: Homo Ludens. Beacon Press, Boston (1955)

Ip, B.: Narrative structures in computer and video games: part 1 – context, definitions, and initial findings. Games and Cult. **6**(2), 103–134 (2011a)

Ip, B.: Narrative structures in computer and video games: part 2 – emotions, structures, and archetypes. Games Cult. **6**(3), 203–244 (2011b)

Jennett, C., Cox, A., Cairns, P., et al.: Measuring and defining the experience of immersion in games. In: International Journal of Human-Computer Studies **66**(9), 641–661 (2008)

Juslin, P.: Communicating emotion in music performance: a review and a theoretical framework. In: Juslin, P., Sloboda, J. (eds.) Music and Emotion: Theory and Research. Oxford, Oxford University Press (2001)

Juul, J.: A Casual Revolution. MIT Press, Cambridge, MA (2010)
Kamp, M.: Musical ecologies in video games. Philos. Technol. **27**(2), 235–249 (2013)
King, G., Krzywinska, T. (eds.): Screenplay: Cinema/Videogames/Interfaces. Wallflower, London (2002)
Klevjer, R.: In defense of cutscenes. In: Proceedings of Computer Games and Digital Cultures Conference. Tampere University Press, Tampere (2002)
Lammes, S.: Spatial regimes of the digital playground: cultural functions of spatial identification in post-colonial computergames. In: Proceedings of Mediaterr@:Gaming Realities. A Challenge for Digital Culture. 236-243 (2006)
Lankoski, P.: Computer games and emotion. In: Sageng, J., Fossheim, H., Larsen, T.M. (eds.) The Philosophy of Computer Games. Springer, Dordrecht (2012)
Lerner, N.: Mario's dynamic leaps: musical innovations (and the spectre of early cinema) in Donkey Kong and Super Mario Bros. In: Donnelly, K.J., Gibbons, W., Lerner, N. (eds.) Music in Video Games: Studying Play. Routledge, New York and London (2014)
Loyer, E.: Parasocial and social player-avatar relationships: social others in Thomas was alone. Press Start. **2**(1), 21–32 (2015)
Madigan, J.: The Psychology of Immersion in Video Games. The Psychology of Video Games. http://www.psychologyofgames.com/2010/07/the-psychology-of-immersion-in-video-games (2010)
Moberley, K.: Preemptive strikes: ludology, narratology, and deterrence in computer game studies. In: Thompson, J., Ouellette, M. (eds.) The Game Culture Reader. Newcastle-upon-Tyne, Cambridge Scholars Publishing (2013)
Nacke, L., Lindley, C.: Affective ludology, flow and immersion in a first-person shooter: measurement of player experience. Loading… **3**(5), 1–21 (2009)
Newman, J.: Videogames. Routledge, London (2004)
O'Grady, D.: Movies in the Gameworld: revisiting the video game cutscene and its temporal implications. In: Ouellette, M., Thompson, J. (eds.) The Game Culture Reader. Cambridge Scholars Publishing, Newcastle upon Tyne (2013)
Perron, B., Schröter, F. (eds.): Video Games and the Mind: Essays on Coginition, Affect and Emotion. McFarland & Company, Jefferson, NC (2016)
Phillips, W.: A Composer's Guide to Game Music. MIT Press, Cambridge, MA (2014)
Phillips, W.: Composing for virtual reality and interactive video game (interview). In: Scoreit Magazine. http://magazine.scoreit.org/interview-winifred-phillips-game-composer (2017)
Rouse, R.: Embrace your limitations: cut-scenes in computer games. Comput. Graph. 7–10. http://dl.acm.org/citation.cfm?id=307713 (1998)
Sageng, J.R., Fossheim, H., Larsen, T.M. (eds.): The Philosophy of Computer Games. Springer, New York (2012)
Salen, K., Zimmerman, E.: Rules of Play. MIT Press, Cambridge, MA (2003)
Seah, M., Cairns, P.: From immersion to addiction in videogames. In: Proceedings of the 22nd British HCI Group Annual (Vol.1) British Computer Society, 55–63 (2008)
Sicart, M.: Digital games as ethical technologies. In: Sageng, J.R., Fossheim, H., Larsen, T.M. (eds.) The Philosophy of Computer Games. Springer, New York (2012)
Smith, J.: Movie Music as Moving Music: Emotion, Cognition and the Film Score. In: Plantinga, C., Smith, G. (eds.) Passionate Views: Film, Cognition and Emotion. The John Hopkins Press, Baltimore (1999)
Stenros, J.: In defence of a magic circle: the social, mental and cultural boundaries of play. In: Transactions of the Digital Games Research Association. Tampere University Press, Tampere (2014)
Stevens, R., Raybould, D.: The reality paradox: authenticity, fidelity and the real in Battlefield 4. The Soundtrack **8**(1), 57–75 (2015)
Summers, T.: Understanding Video Game Music. Cambridge University Press, Cambridge (2016)
Tavinor, G.: Videogames and fictionalism. In: Fossheim, H., Mandt Larsen, T., Sageng, J.R. (eds.) The Philosophy of Computer Games. Springer, New York (2012)

Taylor, T.: Play Between Worlds. Exploring Online Game Culture. MIT Press, Cambridge, MA (2006)
Thon, J.-N.: Narrative comprehension and video game storyworlds. In: Perron, B., Schröter, F. (eds.) Video Games and the Mind: Essays on Cognition, Affect and Emotion. McFarland & Company, Jefferson, NC (2016)
Van Elferen, I.: Analysing Game Musical Immersion: the ALI model. In: Kamp, M., Summers, T., Sweeney, M. (eds.) Ludomusicology: Approaches to Video Game Music. Equinox Publishing, Sheffield (2016)
Warnke, M.: On the spot: the double immersion of virtual reality. In: Liptay, F., Dogramaci, B. (eds.) Immersion in the Visual Arts and Media. Brill, Leiden (2015)
Weiss, M.: How videogames express ideas. In: DiGRA '03 – Proceedings of the 2003 DiGRA International Conference: Level Up (2003)
Whalen, Z.: Play along – an approach to videogame music. Game Stud. Int. J. Comput. Game Res. **4**(1). http://www.gamestudies.org/0401/whalen/ (2004)
Wiebe, E., et al.: Measuring engagement in video game-based environments: Investigation of the User Engagement Scale. Computers in Human Behavior **32**, 123–132 (2014)
Willemsen, S., Kiss, M.: Unsettling melodies: a cognitive approach to incongruent film music. Acta Univ. Sapientiae Film Media Stud. **7**, 169–183 (2013)
Williams, D., et al.: A perceptual and affective evaluation of an affectively-driven engine for video game soundtracking. ACM Comput. Entertain. **14**(3). https://cie.acm.org/articles/perceptual-and-affective-evaluation-affectively-driven-engine-video-game-soundtracking/ (2016)
Wingstedt, J., et al.: Narrative music, visuals and meaning in film. Vis. Commun. **9**(2), 193–210 (2001)
Zehnder, S., Lipscomb, S.: The role of music in video games. In: Vorderer, P., Bryant, J. (eds.) Playing Video Games: Motives, Responses and Consequences. London, Lawrence Erlbaum Associates (2006)
Zhang, J., Fu, X.: The Influence of Background Music of Video Games on Immersion. Journal of Psychology & Psychotherapy **05**(04), 1–7 (2015)

Chapter 11
The Impact of Multichannel Game Audio on the Quality and Enjoyment of Player Experience

Joe Rees-Jones and Damian T. Murphy

Introduction

The term *multichannel audio* is used in reference to a collection of rendering techniques designed to present sound to a listener from multiple directions. In general, the aim of such techniques is to enrich a listener's experience of multimedia content through them feeling a sense of involvement. The ways in which video games are created, and how they are played, makes them well suited to the benefits of multichannel audio. Spatialised sound cues can be used to fully envelop the player in audio, creating immersive virtual sound environments that dynamically react to player input. This chapter focuses on three-dimensional style games, where a large portion of the action can take place off-screen, either behind or to the sides of the player's viewpoint (Collins 2013). Therefore audio cues can be used to influence the player's actions by guiding them towards the next narrative event/objective or warning them of impending threats, potentially reducing the amount of visual information needed on-screen. From this, it is not unreasonable to think that video games enhanced with multichannel audio could lead to a better quality of experience, through making them altogether more engaging and offering clear tactical advantages.

A user's quality of experience (QoE) is considered when it is desirable to know the extent to which some multimedia application is liked, as defined by:

> The degree of delight or annoyance of a person whose experiencing involves an application, service or system. It results from a person's evaluation of the fulfillment of his or her expectations and needs with respect to the utility and/or enjoyment in light of the person's context, personality and current state (Raake and Egger 2014).

J. Rees-Jones (✉) • D.T. Murphy
University of York, York, UK
e-mail: jrj504@york.ac.uk; damian.murphy@york.ac.uk

D. Williams, N. Lee (eds.), *Emotion in Video Game Soundtracking*, International Series on Computer Entertainment and Media Technology,
https://doi.org/10.1007/978-3-319-72272-6_11

Commonly, QoE studies focus on self-reported user responses after the manipulation of one or more pre-determined influencing factors (IFs). With regard to audio related content, the number of audio channels used for presentation is considered to be one of these factors (Reiter et al. 2014). Although not directly related to QoE, an international survey by Goodwin (2009) also suggests that multichannel audio is considered to be important when playing a game. However, beyond these points, there is very little evidence in the literature to more confidently support the claim that multichannel game audio has a significant influence. It is also interesting to think on the reasons for which multichannel game audio could impact player QoE. Is it because a multichannel listening system can offer high spatial quality, thus enhancing the overall auditory aesthetic? Or, could it be that games become easier to play because of multichannel audio? Or is it a combination of the two? This chapter will explore the different ways in which multichannel audio can be used in video games and how its implementation may work to influence the player's perceived quality of experience.

Multichannel Audio in Gaming

Before delving into the potential impact multichannel audio may have on gameplay experiences, it is important to note the current state-of-the-art and reflect on some of the milestones in game audio up to this point. Although audio had been common in arcade and pinball machines for some time, early home gaming hardware had very limited sound capabilities. The which is often regarded as the first home gaming console, didn't have any audio output at all (Magnavox, 2012) and later machines from the mid 70's, such as PONG (Atari 1972), were limited to mono beeps.

Stereo

The Commodore Amiga 1000 (Commodore 1985) is believed to be the first home gaming computer that could output two separate audio channels, offering the possibility for stereo playback. This was thanks to the Paula soundchip (Weske 2015), which could also handle 8-bit digital audio, allowing games to move away from commonly used frequency modulation (FM) synthesis techniques, and instead towards employing recorded samples. Stereo still has the largest user base amongst players and is standard in almost all games (Goodwin 2009). The technique is based on manipulating the relative amplitude of two audio channels to create the impression of movement, whilst also separating sound effects within the stereo image (Rumsey and McCormick 2012). It is important to note that there can be some disparity between visual and audio feedback as the angle between two loudspeakers becomes wider. This may have potential implications on how a user judges an experience, where sounds will be perceived to 'pull' towards the loudspeaker closest to the intended position (Rumsey 2012).

Surround-Sound

One of the shortcomings of stereo is that imaging to the sides and rear becomes difficult due to there being only two, generally frontally positioned, audio channels. Surround-sound was implemented in a handful of titles for the Super Nintendo Entertainment System (SNES) (Nintendo 1991; Lara 2013), making it the first example of a game console to utilise the Dolby Surround home theatre standard (Dolby 1982). The technique extended the conventional stereo format through the addition of a surround channel used to drive loudspeakers to the rear of the listening space (Robjohns 2001). For gaming, this rear channel was often reserved for ambient sounds, such as weather effects, or music (Simpson 2011). Notable games include Jurassic Park (Ocean 1993), King Arthur's World (Argonaut 1993), Vortex (Argonaut 1994) and Samurai Spirits/Shodown (SNK 1993).

The next major development for gaming came with the release of the Sony Playstation 2 (PS2) (Sony 2000) which, with its inclusion of a DVD drive, could output up to 5.1 surround-sound. As illustrated in Fig. 11.1, the standard retains an optimal stereo pair (L and R) but adds two surround channels (LS and RS) used to present audio all around the listening space. In film, the centre channel (C) is usually reserved for dialogue. However, in gaming this channel is often also used for regular sound effects. A subwoofer (not shown in Fig. 11.1) can also added to sweeten low frequency effects (LFE). A particularly noteworthy example was Shadow of the Colossus (ICO 2005), where the surround mix really helps to accentuate the scale of the world and the gigantic creatures within it. The Playstation 3 (PS3) (Sony 2006) introduced 7.1 surround-sound to console gamers, which expanded on 5.1 by including two additional rear surround channels (LBS and RBS) and moving the LS and RS channels slightly forward, as illustrated in Fig. 11.2. It is now the standard multichannel rendering format for the majority of big budget games, including those for the Playstation 4 (PS4) (Sony 2013) and Xbox One (Microsoft 2013). Although surround-sound mixing strategies vary between games, audio is usually rendered to every channel in a surround-sound system (Kerins 2013). The core aim is to give players the impression of a reactive sound scene, that dynamically responds to their physical input.

Game audio designed for surround-sound listening can be presented over any regular stereo system, at the expense of fuller spatialisation. Surround channels are attenuated and then combined with the front left and right to ensure no sound effects are lost, in a process called down-mixing (ITU-R 2004). The method is beneficial for content creators, as it negates the need to generate separate mixes of what is essentially the same audio material. It also provides an easily implemented solution for headphone presentation.

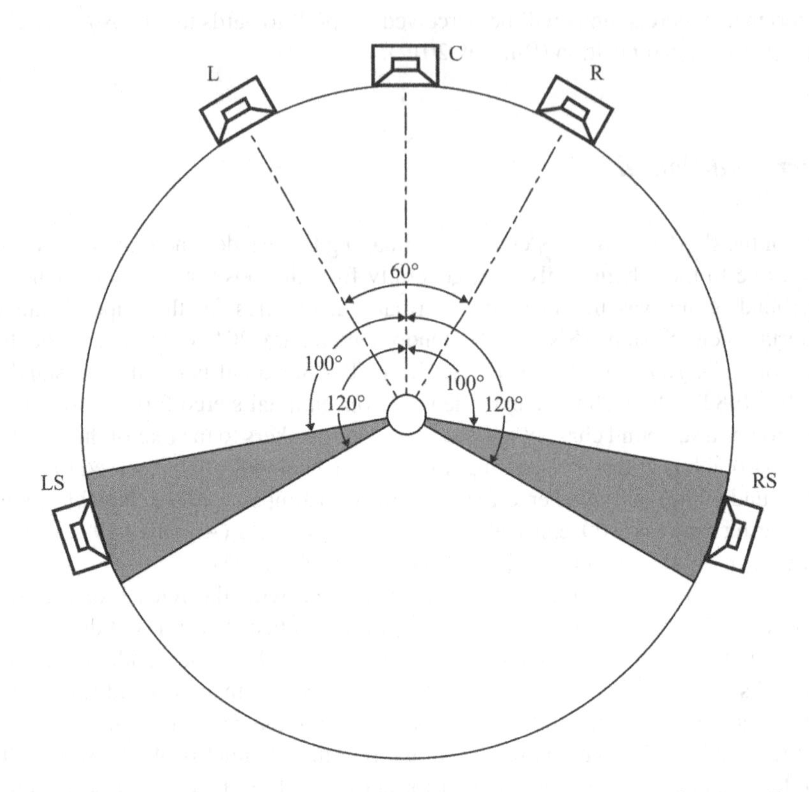

Fig. 11.1 The loudspeaker angles recommended for 5.1 surround-sound listening, as outlined in ITU-R BS 775 (ITU-R 2004). Sounds can be panned around the listener using the left and right surround channels (LS and RS). The arrangement retains an optimally spaced stereo pair (L and R)

Virtual Home Theatre (VHT)

Virtual Home Theatre (VHT) systems offer another headphone based approach for surround-sound listening. Loudspeakers are virtualised by processing individual audio channels with head related transfer functions (HRTFs) (Rumsey 2012). Simply put, these functions encapsulate the way in which a sound wave changes as it travels from the source to the ears, defined by time/phase differences, and filtering effects caused by anatomical features, such as the head and pinna. These changes are decoded by the brain to infer the location of a sound source. For a headphone based VHT system, HRTFs are gathered by taking impulse responses at the ears of a test listener or dummy head relative to loudspeaker positions conforming to either 5.1 or 7.1 surround-sound standards (Horbach and Boone 1999; Aristotel Digenis 2015). Examples in gaming include the Turtle Beach i60 headset (Turtle Beach 2017) and Razer Surround software (Razer 2017).

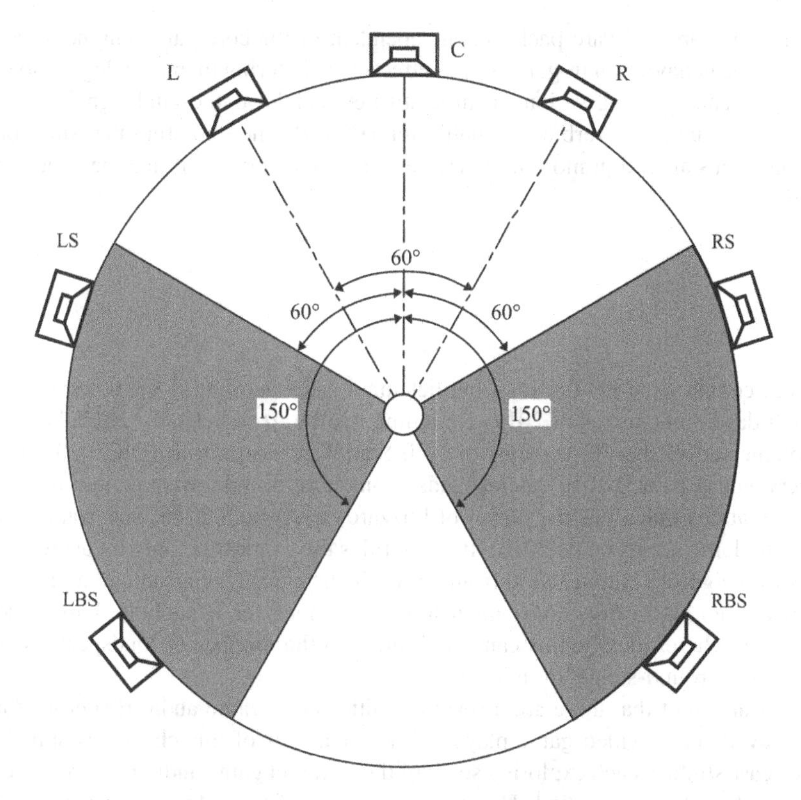

Fig. 11.2 The loudspeaker angles recommended for 7.1 surround-sound listening (ITU-R 2004). It extends 5.1 by pushing LS and RS forward and adding Left Back Surround (LBS) and Right Back Surround (RBS) channels. In a VHT system, these loudspeakers are virtualised for headphone listening

Ambisonics

The term Ambisonics refers to a multichannel encoding technique, developed by Gerzon (1973), based on the decomposition of a sound scene using spherical harmonics. The format offers an interesting alternative to conventional surround-sound, since Ambisonically encoded audio can, in theory, be decoded to any number of loudspeakers in any configuration. Some games do have an Ambisonic audio option, the most notable being DiRT by Codemasters (2007), although it has not been widely adopted. One could speculate that this is because there are not yet any standard sound design workflows for Ambisonic game audio. However as of 2016, plug-ins to encode and decode Ambisonic game audio have been available in Wwise (Audiokinetic 2016), a leading game audio middleware by Audiokinetic. Middlewares such as this are software packages, independent of the core game engine, used to control the behavior on in-game sound effects by triggering them, cataloging source positions and applying various manipulative effects through digital signal processing (DSP), such as reverb and equalisation (EQ). By incorporating the Ambisonic format as a standard in more middlewares and game engines, it may become more common.

Future

It isn't certain what the future of multichannel game audio may be, however some recent developments (as of 2017) are telling. Dolby Atmos (Dolby 2012) has been implemented in the PC versions of both Star Wars Battlefront (DICE 2015) and Overwatch (Blizzard 2016) and expands upon the surround format in multiple ways (Dolby atmos unleashes the power of blizzards overwatch 2016; Star wars battlefront in dolby atmos on pc 2016). In these titles this is most notably by using channels for height, to output sound from above the listener. The introduction of virtual reality (VR) also offers some interesting challenges for headphone based audio playback where, ideally, this can't be limited to the number of loudspeakers in a physical surround-sound arrangement.

It is apparent that there are a several multichannel game audio rendering solutions available to video game players. The remainder of this chapter is split into three case studies, each exploring some of the different game audio rendering methods that have been identified. The studies comprise of several subjective and objective listening tests designed to better understand multichannel audio as an influencing factor on a player's quality of experience.

Percieved Spatial Quality and Player Preferences: Case Study 1

The purpose of this study was to explore how the perceptual qualities of multichannel game audio can influence a player's subjective rating of the overall game session, as quantified through a preference score. This was based on the argument that the spatial characteristics of audio content contribute significantly towards a listener's perception of overall sound quality/fidelity (Letowski 1989; Rumsey et al. 2005; Zielinski et al. 2005). If a player believes a playback system is more able to convey the spatial qualities of the game soundtrack, will that game session be most preferred?

Experimental Methods

Spatial audio attributes are terms used to describe the spatial characteristics of audio. These attributes form the foundation for subjective listening tests with the purpose of determining the spatial quality of a multichannel playback system. Comprehensive lists of spatial attributes, with descriptors, are given in a number of publications concerning the assessment of listening systems. Most notably this includes the Spatial Audio Quality Inventory (SAQI) by Lindau et al. (2014), as well as work by Le Bagousse et al. (2010), Rumsey (2012) and, Bech and Zacharov (2006). For this study, the most frequently mentioned attributes are condensed into a shorter list, with simplified descriptions to ensure understandability over a wide range of participants. Attribute quality was rated on a 5-point numerical scale structured as: (1) Bad, (2) Poor, (3) Fair, (4) Good and (5) Excellent (TIR Assembly 2003). The summarised list of spatial attributes, with descriptors, are given below:

Localisation Accuracy

Refers to how easy it is to identify the direction in which a sound source is originating. There should be good agreement between the visual location of an object/character in the game world and the sound it emits.

Distance Accuracy

Refers to the perceived distance of sound sources. There should be good agreement between a sound source's perceived distance and the position of its related in-game object.

Sense of Depth

Refers to the perceived front-back definition of the sound scene and the sound sources within it. A scene with a good sense of depth will help to create a sense of auditory perspective.

Sense of Width

Refers to the perceived left-right definition of the sound scene and the sound sources within it.

Envelopment

Refers to the extent to which the player feels surrounded by the sound presented in the scene.

User preference ratings give a good indication as to an individual's quality of experience, using the assumption that a more fulfilling experience will be preferred over a less fulfilling one. Choisel and Wickelmaier (2006, 2007) have also shown that there is a relationship between the degree to which a multichannel listening system is preferred and how successful it is at conveying certain auditory attributes/ sensations to a listener. For this work, the preference for one listening condition over another (A over B) was rated using a 7-point paired comparison scale. For each comparison the scale was structured: Strong preference for A, preference for A, slight preference for A, no preference, slight preference for B, preference for B, strong preference for B. This paired comparison design assumes that the preference rating given for one condition will yield the opposite rating for the other condition (Scheffe 1952; David 1963). For example, if the user feels a strong preference towards stimuli A it is assumed that this means there was a strong non-preference for stimuli B.

Mono, stereo and 7.1 surround-sound were the experimental conditions, based on the fact that they are audio rendering standards used in the majority of commercial video game content. It was therefore important to choose a game capable of outputting audio to these chosen playback conditions. *The Last of Us: Remastered* (Sony, 2014), developed by Naughty Dog for the PlayStation 4 (PS4), was used for this reason. Also, the importance of audio is stated early in the game's narrative, where the player is encouraged to listen for potential threats in order to gain a tactical advantage over enemy non-playable characters (NPCs). These crucial sound cues are further emphasised when listening to the game's audio over a 7.1 surround-sound loudspeaker system, significantly influencing the way in which the game can be played. Critically, the game has received praise for its use of audio in the wider game audio community and was at one point the most awarded game in history (DualShockers 2014).

To ensure that playing the game would not overly distract participants from the audio rating task, it was also important to consider those aspects of the content relating to gameplay/interaction. Work by Zielinski et al. (2003) suggests that the visual aspects of a game world, and the attention required to successfully interact with it, can have a significant influence on an individual's ability to rate audio quality. The introductory sequence of *The Last of Us: Remastered* was chosen for ease of playability, in an attempt to not distract participants from the audio rating task. The player is required to follow a fairly simple and linear path with clear instructions from in-game events and sequences (Fig. 11.3). The majority of the audio cues are scripted and will not trigger until the player encounters a particular section, ensuring similar auditory experiences between different players on multiple play-throughs. There are also a limited number of fail-states during the play-through, where, even if the player does fail an objective, they were able to continue with minimal consequence/loading time.

Fig. 11.3 A screenshot from the introductory sequence of *The Last of Us: Remastered* where the player is only required to move through the scene with limited interaction

There was a concern that results might succumb to some bias, due to it being easy to derive each listening condition based on the number of active loudspeakers. Therefore, in an attempt to keep the test blind, all 8 loudspeakers used in a 7.1 surround-sound arrangement were active for all three of the conditions. For the mono condition, a mix-down to mono of the game audio was output at an equal loudness from all of the loudspeakers. This is what is often referred to as 'full', or 'big', mono. For the stereo condition, audio intended for the left channel was output from the three loudspeakers of a 7.1 arrangement positioned to the left of the listener. The same was done with the right channel but to the right-hand loudspeakers. The centre loudspeaker output a sum of the stereo channels. Although presenting the game audio in this way meant sound would be output from all around the listener, it was expected that the spatial quality would still be limited. The objective limitations of mono and stereo means that discrete auditory cues to the rear or sides of the listener aren't possible, instead, this information is mixed in with the respective front channels. For this reason, a separate control group, comparing regularly configured stereo (i.e. only two loudspeakers placed at ±30°) with 7.1 surround-sound, was also included.

Discussion

Results suggested that a listening condition that was perceived to have high spatial quality was also most preferred by participants when engaged in the game. This was especially clear in regards to the mono condition, which consistently received the lowest spatial quality ratings and was also least preferred. However, both the stereo and 7.1 surround-sound conditions received similarly high spatial quality ratings and neither was preferred significantly more than the other. This result was surprising, as it was expected that, due to the high potential for spatialisation, 7.1 surround-sound would be perceived to have higher spatial quality than that of stereo. The control group's results were more obvious, where the spatial quality of regularly configured stereo was perceived to be much lower than that of 7.1 surround-sound and was also not preferred by the majority of participants.

Based on the outcome of the control group, it is likely that participants found it difficult to distinguish 'big' stereo from 7.1 surround-sound. The panning between the left and right channel will have been exaggerated to some extent by the fact that the they were output from all the loudspeakers of a 7.1 arrangement. This extreme panning may have been perceived by listeners to be more spatial than regular stereo, resulting in an overall more positive spatial quality rating. It is important to note that 'big stereo' will naturally feel enveloping, since audio is physically output from all around the listener, even if the spatial information is not completely accurate in respect to the games visual feedback. It is well regarded that visual stimuli can have significant effects on the perception of spatial attributes of audio stimuli, especially in regard to sound source localisation (Moore 2012; Werner et al. 2012). This was apparent in several participants' comments. It was implied by one individual that they were able to localise the sounds of two passing police cars, even when using 'big mono'. Another stated they 'could hear a helicopter passing over their head', even though loudspeakers mounted above the listener were not used in any of the three listening conditions.

Although it was unexpected that 'big stereo' was considered to have similarly high spatial quality to 7.1 surround-sound, it is important that both were equally preferred. From this it can at least be inferred that playing a game in a perceptually more spatial listening environment will be preferred, and potentially offer a more fulfilling gaming experience. This has positive implications for those gamers who cannot invest in a full surroundsound system, where 'big stereo' could be a viable alternative. Rather than using the eight loudspeakers needed for 7.1 surround-sound, perhaps players could benefit from more compact systems capable of outputting 'big stereo', providing a heightened sense of spatialisation and envelopment over regular stereo.

Headphone Based Audio Rendering and Player Preferences: Case Study 2

It is often difficult to convince users to commit to a full 7.1 surround-sound system, due to both the costs and the space needed for multiple loudspeakers. Also, the consistency of loudspeaker placement between different listening spaces (i.e. living rooms) is questionable (Goodwin 2009). A more practical approach is to virtualise the loudspeaker positions of a surround-sound system for listening over a pair of stereo headphones. For this study, headphone based equivalents of stereo and 7.1 surround-sound were compared. The aim is to find whether a VHT rendering of 7.1 surround-sound offered a significantly different experience compared to a stereo down-mix. Again, this was based on a comparison of the perceived spatial quality and preference ratings between the two conditions.

Previous Work on VHT/Stereo Comparisons

Zacharov and Lorho (2004) conducted a study comparing a number of virtual 5.1 surround-sound systems with stereo down-mixes of the same audio material. A range of audio stimuli were used, including a video game. However subjects did not directly interact with the stimuli during the test. None of the virtual surround-sound methods were found to out-perform the downmix, and in some cases the down-mix was slightly preferred. A more recent study done at the BBC by Pike and Melchior (2013) yielded comparable results where virtual 5.1 surround sound was perceived to have similar overall quality to a stereo down-mix anchor. Sousa (2011) investigated the subjective spatial quality of a 16-channel Ambisonic system rendered binaurally. Although the binaural system was considered to have enhanced spatial quality, a stereo down-mix was still preferable among subjects.

The conclusions drawn from these studies do not provide positive arguments for the use of virtual surround-sound systems, however, in none of these cases were participants asked to interact with audio stimuli through playing a game. It is therefore of interest to investigate whether similar comparisons can be made between virtualised and down-mixed multichannel game audio, with the added variable of interactivity.

Experimental Methods

Natively, *The Last of Us: Remastered* does not have a VHT version of the surround mix. For this reason, the VHT headphone output was generated in Max/MSP (Cycling 2017), by processing the individual audio channels of the PS4 with generic HRTF measurements found in the Spat library of objects by IRCAM (Jot and Warusfel 1995). The down-mix to stereo was also done in Max/MSP, by attenuating the surround channels and summing them with the left and right, according to ITU-R BS.775 (ITU-R 2004). Both conditions were presented over a pair of Beyerdynamic DT 990 stereo headphones. All participants played the game using both the VHT system and the down-mix, in a randomised order. After each game session, spatial quality was rated as outline in case study 1. A preference score was given after the completion of both playthroughs.

Discussion

On the whole, results suggested that the virtual surround-sound condition was not preferred over the stereo down-mix, and there was no significant improvement in the perceived spatial quality (Fig. 11.4). Like with the first case study, the spatial attributes were rated relatively highly for both conditions, again suggesting that high spatial quality was preferable (Fig. 11.5). This outcome was not entirely unexpected, as similar work presented by Pike and Melchior (2013), Zacharov and Lorho (2004), and Sousa (2011) yielded similar results. The fact that a video game was used in place of more traditional non-interactive material, does not seem to have had any significant impact on the perceived performance of the 7.1 system. It is therefore difficult to conclude whether in using a VHT system, a player's game experience can actually be improved in comparison to a standard/down-mixed stereo rendering.

Ideally, unique head-related transfer function data-sets should be used for each participant as part of a VHT, with some simulation of room acoustics to improve externalisation (Rumsey 2012). This may have resulted in the lack of perceptual difference between the conditions. The use of non-individualised HRTFs can significantly reduce an individual's ability to perceive spatialised aspects of audio material when listening over headphones (Begault and Trejo 2000). In extreme situations this can result in the incorrect spatialisation of audio and undesirable timbral colourations. Unfortunately the use of generic/non-individualised HRTFs is representative of virtual surround-sound gaming headsets, due to the difficulties in collecting individualised measurements.

It is also difficult to say whether the multimodal task of playing a video game served to have a positive or negative influence on the way in which participants were able to rate the listening conditions. As work by Zeilinksi suggests, subjects may not be able to effectively rate audio content with interactive and visual elements

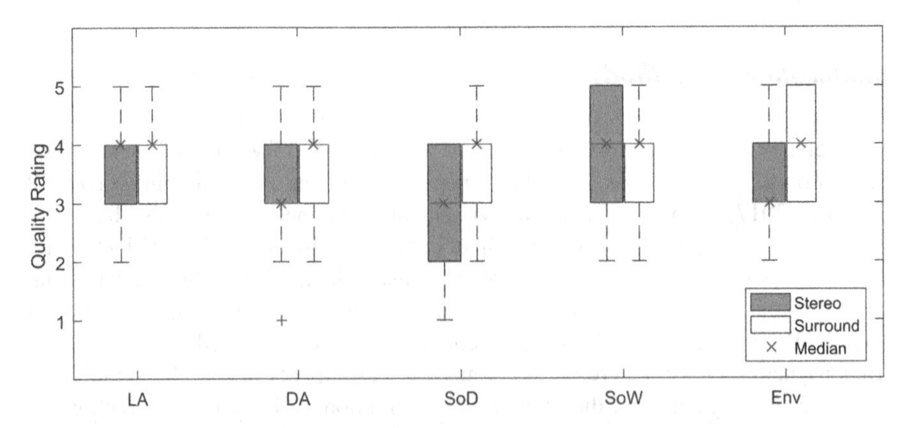

Fig. 11.4 Spatial quality scores for stereo and 7.1 surround-sound presented over headphones. Both were perceived to have high spatial quality, however there was no significant difference between the two

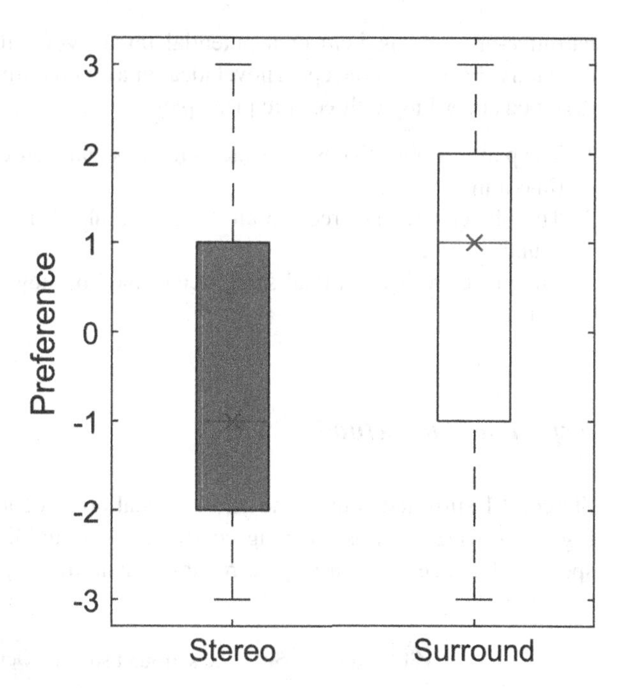

Fig. 11.5 Preference ratings for the headphone based stereo and 7.1 listening conditions. The sparse distribution of ratings suggests that neither listening condition was preferred significantly more than the other

(Zielinski et al. 2003). In regard to the two studies already presented here, participants had to play a relatively long section of the game (12–16 min) before they had a chance to give their opinions on the audio presentation. It is therefore not unreasonable to surmise that results, in both cases, may have succumbed to some bias, based on a listener's inability to only focus on the auditory aspects of the game session. It may be that listening tests based on subjective measures are not appropriate in experimental situations where participants are so engaged in the task, and instead, objective quantification might prove to be more reliable. This could include quantifying a player's score, or the time it takes to complete an ingame task. Data can then be collected in real-time, potentially reducing the cognitive load on participants.

The Impact of Multichannel Audio on Player Performance: Case Study 3

The aim of the final study in this series is to explore the potential of an objectively measurable metric as a way to quantify player performance. The test is designed to determine whether spatialised game audio can improve player performance in a gamified localisation task, and thus influence the overall game experience. If a player is able to better determine the location of an in-game event, such as a narrative-progressing item or a potential threat, will this inform their gameplay decisions? From this concept, a novel idea for a comparative listening test is derived, designed according to three core principles:

1. The player's objective is to locate as many sound sources as possiblein the given time limit.
2. The player does not receive any visual feedback regarding the positionof the sound source.
3. The player receives a final score determined by how many sound sourceswere found.

Experimental Methods

Stereo, 7.1 surround-sound, and an octagonal array of loudspeakers, as shown in Fig. 11.6, were used as listening conditions. An equidistant array of eight loudspeakers has been shown to give reliable phantom imaging all around a listener,

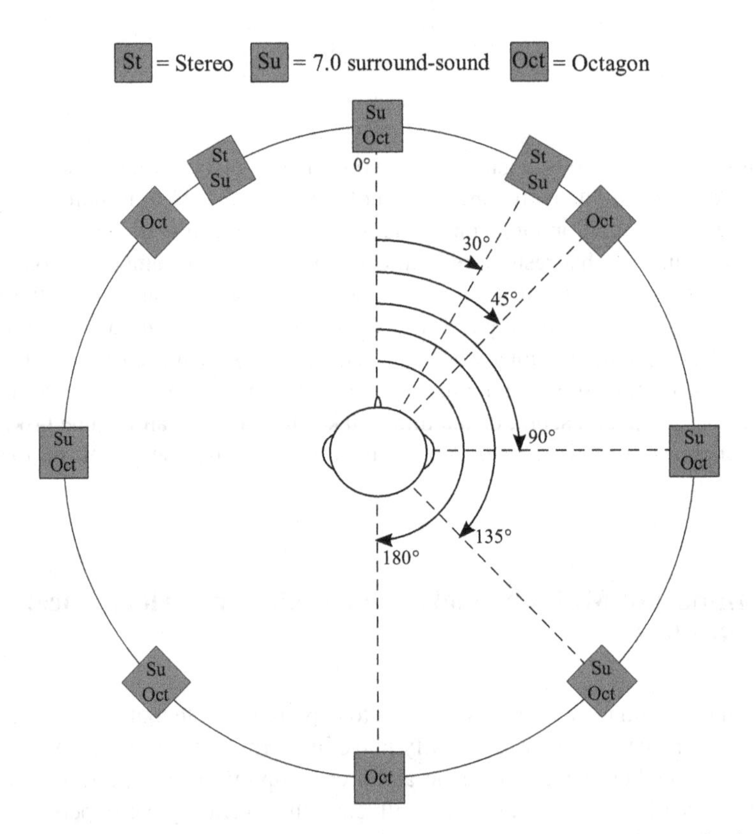

Fig. 11.6 Loudspeaker angles for stereo, 7.1 surround-sound and an octagonal array. Angles were symmetrical to the left and right of a forward-facing listener

potentially improving source localisation (Martin et al. 1999). This gives an improvement on both stereo and 7.1 surround-sound, where the inconsistent placement of loudspeakers leads to sound image instability laterally and to the rear (Theile and Plenge 1977). It was therefore expected that player's would be able to more easily locate the sound source when using an octagonal array, thus giving them a higher score. However, the loudspeakers in front of the listener need to be spaced at a wider angle than those in stereo and 7.1 configurations. Therefore the trade-off between more consistent imaging all around the listener, and the potential for higher resolution frontal imaging, was also of interest.

A custom game environment was needed to realise the aims of the study, as the underlying mechanics/systems were more easily controlled. The game systems were designed using the Unity game engine (Unity 2005), and audio rendering was done in Max/MSP (Cycling 2017). A single sound source was used in the game, the position of which changed as soon as it was successfully located by the player. This was determined by pressing the ‘x’ button on a PS4 controller. If, upon pressing the ‘x’ button, the player-avatar was within the radius of the sound source, the source would move to a random new location at least 10 m away, and their score increased by 1 (Fig. 11.7). Random positioning was implemented to ensure that players could not learn sound source positions after playing the game multiple times. If the ‘x’ button was pressed and the player was not within the radius of the sound source then the current position was maintained with no increase in score.

Players were asked to locate a transient sine tone at a frequency of 440 Hz, repeating every half a second. A short delay effect was also applied to the tone, giving a sonar-like effect. The amplitude of the sound changed dynamically depending

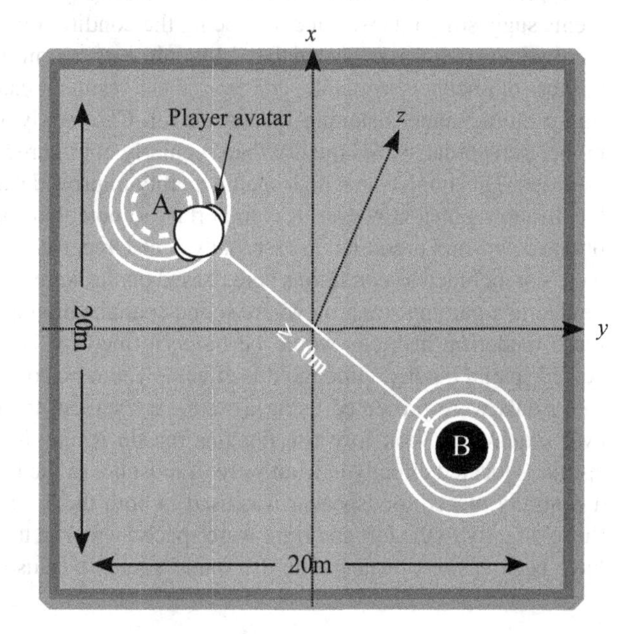

Fig. 11.7 A conceptual illustration of a player correctly locating the sound source in its current position (**A**) by entering its radius and pressing ‘x’ on the gamepad. The sound source then moves to a random new position (**B**) at least 10 m away from the player's current position

on the distance from the player. As the player moved towards the source the amplitude increased, and as they moved further away it decreased, providing a simulation of distance attenuation. An ascending sequence of notes were played upon every correct localisation to give some auditory feedback in-line with the increase in score. Other sound effects such as footsteps, ambience and music were not included as it was decided they might distract from the localisation task.

Before the formal test began, participants were asked to complete a training session, based on a simplified version of the game, allowing them to become familiar with the control scheme and task. For the test, the game was played by participants in all three of the listening conditions in a counterbalanced order. They were not made aware of the conditions prior to, or during, the test. Each game session lasted 2 min 30 s, and a 'Game Over' message, with the player's final score, signified the end of the session. This final score represented the number of correct localisations in a single game session, and was output as a separate.txt file for analysis. Once a subject had been exposed to all of the rendering methods, they were asked to state which of the three conditions they preferred and provide any comments.

Discussion

Players had greater success in the game when listening to audio over a 7.1 surround-sound loudspeaker system, compared to using either stereo or an octagonal array (Fig. 11.8). Localisation scores were consistently higher, where 60% of participants received their highest score when using surround-sound. 7.1 was also the most preferred of the three listening conditions (Fig. 11.9), with participant comments suggesting this was due to it being the condition in which the highest scores were achieved. It was expected that 7.1 would outperform stereo due to the increased number of channels available, and from these results it can be said that players did find a multichannel listening system useful. Case study 1 suggested that 7.1 had higher perceptual spatial quality than regularly configured stereo, which were both assessed. This implies that high spatial quality improved the user's experience of the localisation game. However, it is important to note that 'big' stereo, and the headphone down-mix presented in case study 2, also received high perceptual ratings but were not included as conditions here. Based on the stereo localisation scores, it may be that in a gaming situation where sound spatialisation is integral to the objective, these rendering methods would be outperformed by 7.1, regardless of their perceived spatial quality, although this effect is yet to be confirmed.

Visuals were presented using a stationary screen, therefore players were only ever required to look forward. For this reason it may have been that those loudspeakers located directly in front were of most use in the localisation task. Although a centrally placed loudspeaker was used in both the 7.1 and octagonal conditions, those directly to the left and right were spaced wider in the latter. Therefore, it may have been easier to triangulate the sound source because of the increased frontal

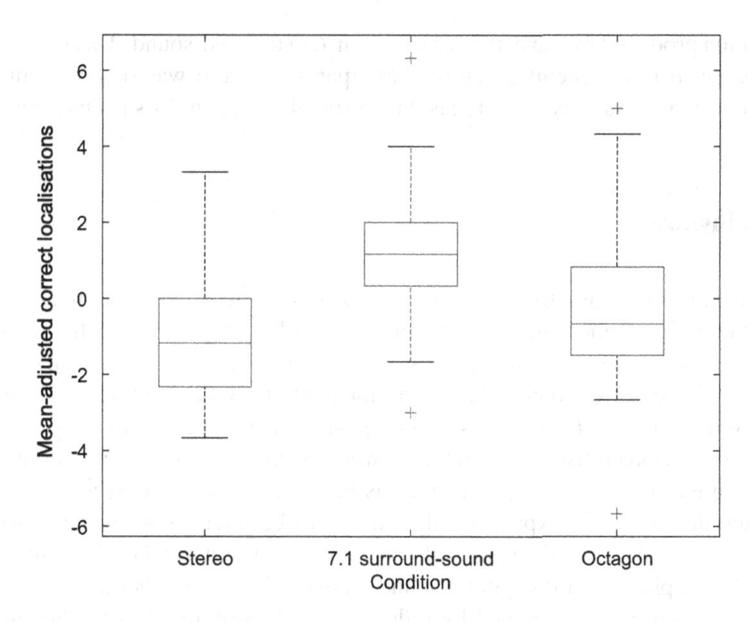

Fig. 11.8 The distribution of player scores for the three listening conditions: stereo, 7.1 surround sound and octagon. The analysis suggested that the highest scores were achieved whilst using 7.1 surround-sound

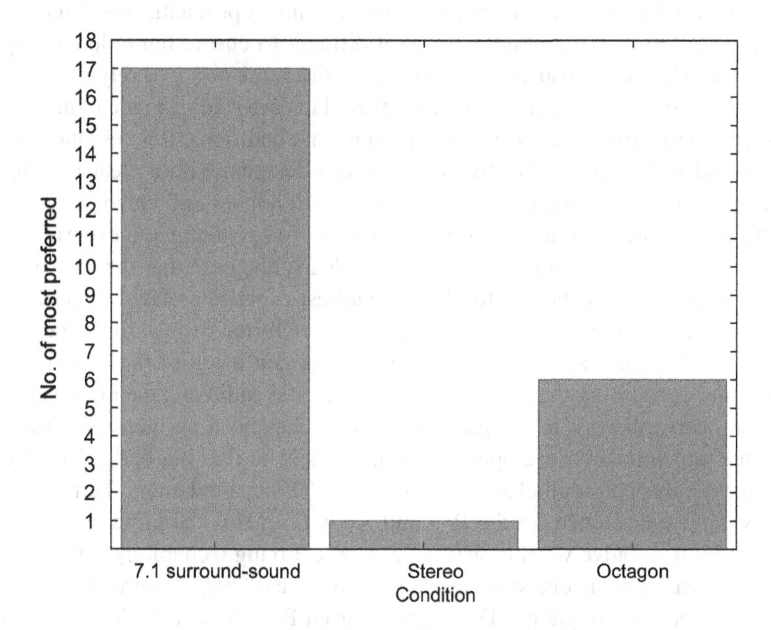

Fig. 11.9 Preference ratings given for each listening condition. 7.1 surroundsound was consistently preferred more often in comparison to the stereo and octagon conditions

resolution produced by the narrower angles in 7.1 surround-sound. This observation was reflected in comments given by participants, where it was stated on multiple occasions that it was easiest to focus on the sound source in 7.1 surround-sound.

Conclusions

The conclusions drawn from these three case studies do suggest that multichannel game audio has some influence on a player's quality of experience. It was shown that perceptually high spatial quality was preferable, even though it was unexpected that stereo and 7.1 surround-sound were rated similarly. Also, player performance improved as a result of using 7.1 surround-sound in a gamified localisation task. The differences between listening conditions were much clearer in the third case study, implying that multichannel game audio is beneficial when used as a tool to drive in-game decisions. To expand on this, it would be interesting to more formally assess the competitive advantages of surround-sound in multiplayer gaming situations. Will a player using surround-sound more easily beat another using stereo?

Lessons can also be learned by reflecting on the different test methodologies. From the first two case studies it became apparent that gathering subjective data may not have been suitable. The attention needed to play the game for the required amount of time will have made it difficult for participants to fully focus on the audio rating tasks. Also, using a pre-existing game became a problem, where the lack of control of the underlying systems made it difficult to ensure multiple participants would have similar exposures. By simplifying the game task and relying on objectively gathered data, the methodology employed in case study 3 made it much easier to interpret any differences between experimental conditions. It would therefore be recommended that more objectively driven test designs, with a clear task for the player, are more appropriate for any experimentation involving video games.

The purpose of the studies in this series is not to say that game experiences are lacking without multichannel audio, as it is clearly the case that the vast majority gamers have their expectations fulfilled, regardless of whether they own a surround-sound system. In fact, through browsing support forums for various games and consoles, it becomes apparent that there is some confusion in the casual gaming community concerning the purposes of multichannel audio and the options that are available. Nevertheless, in the past decade there has been an increasing trend for hardware and software developers (especially those in the 'big budget' market) to implement enhanced multichannel audio, with 7.1 surround-sound being standard for any games developed for the PS4 and Xbox One. It is therefore useful to consider the more creative ways in which this ever evolving technology can be used in a way to drive in-game decisions, as suggested in case study 3, rather than just as a tool to enhance audio quality. The highly praised Papa Sangre 2 (Somethin' 2013) is a game that does this, by making it obvious to the player that only auditory cues will be provided, with no visual feedback. There are clearly many creative avenues to be explored if the role of multichannel game audio as an influence on player quality of experience is to be more fully understood.

References

Argonaut: King Arthur's World: Video Game. Argonaut Games PLC., London (1993)
Argonaut: Vortex: Video Game. Argonaut Games PLC., London (1994)
Aristotel Digenis: Challenges of the headphone mix in games. In: Audio Engineering Society Conference: 56th International Conference: Audio for Games. Audio Engineering Society (2015)
Atari: Pong. Video Game Console. Atari Inc., New York (1972)
Audiokinetic: Ambisonics in wwise: Overview. https://www.audiokinetic.com/products/ambisonics-in-wwise/ (2016). Accessed 20 Sept 2017
Bech, S., Zacharov, N.: Perceptual audio evaluation. Appendix B, pp. 363–366. Wiley (2006)
Begault, D.R., Trejo, L.J.: 3-D Sound for Virtual Reality and Multimedia. AP Professional, Buffalo, NY (2000)
Blizzard: Overwatch. Video Game. Blizzard Entertainment, Inc., Irvine (2016)
Choisel, S., Wickelmaier, F.: Evaluation of multichannel reproduced sound: Scaling auditory attributes underlying listener preference. J. Acoust. Soc. Am. **121**(1), 388–400 (2007)
Choisel, S., Wickelmaier, F.: Relating auditory attributes of multichannel reproduced sound to preference and to physical parameters. In: Audio Engineering Society Convention 120. Audio Engineering Society (2006)
Codemasters: Colin McRae: DiRT. Video Game. Codemasters Software Ltd., Southam (2007)
Collins, K.: Playing with Sound: A Theory of Interacting with Sound and Music in Video Games. MIT Press, Cambridge, MA (2013)
Commodore: Amiga 1000: Video Game Console. Commodore Int., Ltd., Toronto (1985)
Cycling: Max/msp. Computer software, Cycling '74, https://cycling74.com/products/max/ (2017). Accessed 20 Sept 2017
David, H.A.: The Method of Paired Comparisons, vol. 12. C. Griffin, London (1963)
DICE: Star Wars Battlefront. Video Game, EA DICE AB (2015)
Dolby: Dolby Surround. Surround-Sound Format. Dolby Laboratories, Inc., San Francisco (1982)
Dolby: Dolby Atmos: Surround-Sound Format. Dolby Laboratories, Inc., London (2012)
DualShockers: Naughty dog's the last of us is the most awarded game in recorded history by critics. http://www.dualshockers.com/the-last-of-us-is-the-most-awarded-game-in-history-by-critics/ (2014). Accessed 21 Aug 2017
Gerzon, M.: Periphony: with-height sound reproduction. J. Audio Eng. Soc. **21**(1), 2–10 (1973)
Goodwin S.N.: How players listen. In: Audio Engineering Society Conference: 35th International Conference: Audio for Games. Audio Engineering Society (2009)
Horbach, U., Boone M.M.: Future transmission and rendering formats for multichannel sound. In: Audio Engineering Society Conference: 16th International Conference: Spatial Sound Reproduction. Audio Engineering Society (1999)
ICO: Shadow of the Colossus: Video Game. Team ICO, Tokyo (2005)
ITU-R: 775-2, multi-channel stereophonic sound system with or without accompanying picture, int. Telecommunications Union Radiocomunication Assembly, Geneva (2004)
Jot, J.-M., Warusfel, O.: Spat: a spatial processor for musicians and sound engineers. In: CIARM: International Conference on Acoustics and Musical Research (1995)
Kerins, M.: Multichannel gaming and the aesthetics of interactive surround. In: Gorbman, C., Richardson, J., Vernallis, C. (eds.) The Oxford Handbook of New Audiovisual Aesthetics. Oxford University Press, New York (2013)
Lara, C.: Throwback Thursday: Top 5 uses of surround sound on snes. https://www.gamnesia.com/articles/throwback-thursday-top-5-uses-of-surround-sound-on-snes (2013). Accessed 19 Sept 2017
Le Bagousse, S., Paquier, M., Colomes, C.: Families of sound attributes for assessment of spatial audio. In: 129th AES Convention, pages Convention–Paper (2010)
Letowski, T.: Sound quality assessment: concepts and criteria. In: Audio Engineering Society Convention 87. Audio Engineering Society (1989)

Lindau, A., Erbes, V., Lepa, S., Maempel, H.-J., Brinkman, F., Weinzierl, S.: A spatial audio quality inventory (saqi). Acta Acust United Acust. **100**(5), 984–994 (2014)
Magnavox: The magnavox odyssey. http://www.magnavox-odyssey.com/ (2012). Accessed 19 Sept 2017
Martin, G., Woszczyk, W., Corey, J., Quesnel, R.: Controlling phantom image focus in a multichannel reproduction system. In: Audio Engineering Society Convention 107. Audio Engineering Society (1999)
Microsoft: Xbox One: Video Game Console. Microsoft Corp., Albuquerque (2013)
Moore, B.C.J.: An Introduction to the Psychology of Hearing. Brill (2012)
Nintendo: Super Nintendo Entertainment System (SNES): Video Game Console. Nintendo Co., Ltd., Kyoto (1991)
Ocean: Jurassic Park: Video Game. Ocean Software Ltd., Manchester (1993)
Pike, C., Melchior, F.: An assessment of virtual surround sound systems for headphone listening of 5.1 multichannel audio. In: Audio Engineering Society Convention 134. Audio Engineering Society (2013)
Raake, A., Egger, S.: Quality and quality of experience. In: Quality of Experience, pp. 11–33. Springer (2014)
Razer: Razer surround. https://www.razerzone.com/gb-en/surround (2017). Accessed 4 Sept 2017
Reiter, U., Brunnström, K., De Moor, K., Larabi, M.-C., Pereira, M., Pinheiro, A., You, J., Zgank, A.: Factors influencing quality of experience. In: Quality of Experience, pp. 55–72. Springer (2014)
Robjohns, H.: Surround sound explained: Part 2. Sound on Sound **16**(11), 170–176 (2001). [Magazine]
Rumsey, F.: Spatial Audio. CRC Press (2012)
Rumsey, F., McCormick, T.: Sound and Recording: An Introduction. CRC Press (2012)
Rumsey, F., Zieliński, S., Kassier, R., Bech, S.: On the relative importance of spatial and timbral fidelities in judgments of degraded multichannel audio quality. J. Acoust. Soc. Am. **118**(2), 968–976 (2005)
Scheffe, H.: An analysis of variance for paired comparisons. J. Am. Stat. Assoc. **47**(259), 381–400 (1952)
Shadow of the Colossus. https://www.playstation.com/en-us/games/shadow-of-the-colossus-ps2/ (2005). Accessed 19 Sept 2017
Simpson, M.: The first king arthur, and dolby surround. http://www.martinpsimpson.com/2011/09/first-king-arthur-and-dolby-surround.html (2011). Accessed 16 Sept 2017
Somethin': Papa Sangre 2. Video game, Somethin' Else. (2013)
SNK: Samurai Shodown: Video Game. SNK Corp., Suita (1993)
Sony: Playstation 2 (PS2): Video Game Console. Sony Computer Entertainment, Tokyo (2000)
Sony: Playstation 3 (PS3): Video Game Console. Sony Computer Entertainment, Tokyo (2006)
Sony: Playstation 4 (PS4): Video Game Console. Sony Computer Entertainment, Tokyo (2013)
Sony: The Last of Us: Remastered. Video game, Naughty Dog, LLC (2014)
Sousa, F.W.J.: Subjective comparison between stereo and binaural processing from b-format ambisonic raw audio material. In: Audio Engineering Society Convention 130. Audio Engineering Society (2011)
Theile, G., Plenge, G.: Localization of lateral phantom sources. J. Audio Eng. Soc. **25**(4), 196–200 (1977)
TIR Assembly: Itu-r bs. 1284–1: En-general methods for the subjective assessment of sound quality. Technical report, Technical Report, ITU (2003)
Turtle Beach: Ear force i60. https://shop.turtlebeach.com/us/i60 (2017). Accessed 4 Sept 2017
Unity: Unity game engine. Video game engine, Unity Technologies SF, https://unity3d.com/ (2005). Accessed 20 Sept 2017
Werner, S., Liebetrau, J., Sporer, T.: Audio-visual discrepancy and the influence on vertical sound source localization. In: Quality of Multimedia Experience (QoMEX), 2012 Fourth International Workshop on, pp. 133–139. IEEE (2012)

Weske, J.: Digital sound and music in computer games. http://3daudio.info/gamesound/history.html (2015). Accessed 19Sept 2017

Zacharov, N., Lorho, G.: Subjective evaluation of virtual home theatre sound systems for loudspeakers and headphones. In: Audio Engineering Society Convention 116. Audio Engineering Society (2004)

Zielinski, S.K., Rumsey, F., Bech, S., De Bruyn, B., Kassier, R.: Computer games and multichannel audio quality-the effect of division of attention between auditory and visual modalities. In: Audio Engineering Society Conference: 24th International Conference: Multichannel Audio, The New Reality. Audio Engineering Society (2003)

Zielinski, S.K., Rumsey, F., Kassier, R., Bech, S.: Comparison of basic audio quality and timbral and spatial fidelity changes caused by limitation of bandwidth and by down-mix algorithms in 5.1 surround audio systems. J. Audio Eng. Soc. **53**(3), 174–192 (2005)

Chapter 12
Concluding Remarks

Duncan Williams and Newton Lee

At Disney Online, Newton Lee co-developed over 100 online games, many of which employed advanced physics, artificial intelligence, 3D graphics, digitized sound, and original musical scores (Lee and Madej 2012). Guest feedback received by the Disney producers was not all about gameplay but also the appreciation of music in the games.

The power of music to manipulate or augment the listener's emotions is well-known, but the added complication when creating music for videogames is that the emotional characteristic of the music must be congruent with a non-linear, player-driven narrative. By this stage of the book, you are hopefully feeling a mixture of enthused, inspired, or perhaps bewildered. Over the course of these selections it would, of course, be impossible for us to really predict the future, but it is clear that the huge variety of work involving new types of game design, new interfaces for players, virtual reality environments, and most of all, emotionally targeted soundtracking, should create a new type of gaming environment which is richer and more emotionally engaging than we might possibly be able to envision at the time of writing. We predict that in the next ten years the videogame soundtracking landscape may be unrecognizable from the current state-of-the-art.

D. Williams (✉)
Digital Creativity Labs, University of York, York, UK
e-mail: duncan.williams@york.ac.uk

N. Lee
Department of Applied Computer Science, Woodbury University, Burbank, CA, USA

D. Williams, N. Lee (eds.), *Emotion in Video Game Soundtracking*, International Series on Computer Entertainment and Media Technology,
https://doi.org/10.1007/978-3-319-72272-6_12

Evaluation Strategies

For those of you who are interested in soundtrack design, or games industry design, there remain a few areas which have only been touched on briefly in these pages. Specifically, the selection of evaluation strategies is a topic which might merit a book in its own right. It would be remiss not to provide a short summary of the problem, moving forward.

> To listen to data… can be a surprising new experience with diverse applications ranging from novel interfaces… to data analysis problems. (Toharia et al. 2014)

Creativity in soundtracking often becomes a function of the chosen mapping scheme, whereby deliberate specification of data mapping to auditory events provides opportunities to creative expression. Thus, such techniques can be used as part of the soundtrack creation process, if mapping strategies are carefully designed with specific player/narrative outcomes in mind. However, the creative decision making process involved in designing these mapping strategies can by its nature compromise the presentation of the soundtrack in terms of accuracy, and perhaps in terms of overall emotional congruence. Criteria and methodologies for evaluation of soundtracking and auditory display vary (Degara et al. 2013; Hermann et al. 2011; Vogt et al. 2013), and indeed evaluations of mapping strategies are not often performed (Dubus and Bresin 2013). The lack of a universal method for evaluation is a contributor to this problem (Ibrahim et al. 2011) but systems which do not directly require utility (creative approaches) have no explicit need of evaluation other than that of the designers' own aesthetic goals. One might just as well ask the player "Did you like the soundtrack?" We consider the utility of such techniques as a parameter for the evaluation of mapping strategies as comprised of a number of perceptual characteristics related to gameplay but more specific than "like" alone, including immersion, emotional congruence, intuitivity and so on. In this regard, strategies such as multi-criteria decision aid analysis might be the most useful way forward for evaluation of soundtracking decisions. This area remains an exciting and challenging area for further work developing practical and creative systems, perhaps working towards the goal of a soundtracking system which might satisfy the criteria for 'pleasant' listening experience and mediate the potential confound of individual musical preference whilst still offering the full range of options necessary for a specific game or narrative event.

We conclude this book with a quote from the music legend Quincy Jones: "Music has a powerful effect on everybody: children, adults, animals, plants, everything. It's, I think, one of the abstract miracles and phenomena." (Lee 2014).

References

Degara, N., Nagel, F., Hermann, T.: SonEX: an evaluation exchange framework for reproducible sonification. In: Proceedings of the 19th International Conference on Auditory Displays (2013)
Dubus, G., Bresin, R.: A systematic review of mapping strategies for the sonification of physical quantities. Edited by Michael J Proulx. PLoS One. **8**, e82491 (2013). https://doi.org/10.1371/journal.pone.0082491
Hermann, T., Hunt, A., Neuhoff, J.G.: The Sonification Handbook. Logos, Berlin (2011)
Ibrahim, A.A.A., Yassin, F.M., Sura, S., MacDonell Andrias, R.: Overview of design issues and evaluation of sonification applications. In: IEEE, pp. 77–82 (2011). https://doi.org/10.1109/iUSEr.2011.6150541
Lee, N. (ed.): Digital Da Vinci: Computers in Music. Springer, New York (2014)
Lee, N., Madej, K.: Disney Stories: Getting to Digital. Springer, New York (2012)
Toharia, P., Morales, J., Juan, O., Fernaud, I., Rodríguez, A., DeFelipe, J.: Musical representation of dendritic spine distribution: a new exploratory tool. Neuroinformatics. **12**(2), 341–353 (2014). https://doi.org/10.1007/s12021-013-9195-0
Vogt, K., Goudarzi, V., Parncutt, R.: Empirical aesthetic evaluation of sonifications. In: Proceedings of the 19th International Conference on Auditory Display (ICAD2013), Lodz, Poland, 6–9 July 2013

Printed in the United States
By Bookmasters